Table of Contents

Chapter 6 : Coupling, Shaft & Propulsion System 156

Chapter 7 : Lower Unit & Stern-Drives 177

Chapter 8 : Transmission Bearings & Seals 213

Inboard Transmissions & Lower Units

Copyright @ 2012
by
Marine Technical Training

By: Alvaro Lopez
Edition 2014

Marine Technical Training

It is a company dedicated to training Engineers and Marine Technicians in areas such as

- **Marine Electricity**
- **Marine Electronics**
- **Air Conditioning**
- **Diesel Engines**
- **Gasoline Engines**
- **Auxiliary Systems**
  - Hydraulics
  - Steering Systems
  - Trim Tabs
  - Bow Thrusters
  - Water makers

Copyrights

ABYC Certification

This book was designed like a consult handbook for Service, installation and diagnosis of inboard and outboard transmissions in pleasure yachts. Also, it can be used as a textbook for marine engineering courses and as a complementary study guide in order to take the **ABYC Gasoline and ABYC Diesel Certifications**

Recognition

This book is the result of many years of studio and research about marine transmissions and propulsion systems. Most of the items presented here taken as reference, research and articles from other authors. All they want to express my gratitude because thanks to these works, this book will be useful for owners of boats and Marine engineers interested in transmissions service. In the same way some graphics are taken from public internet sites. To the people who have uploaded these images to the cloud I thank its contribution

Chapter 1
Hybrid Configurations

Hybrid Systems

A hybrid boat can achieve propulsion using a fuelled power source (e.g. a diesel or Gas engine) **or** through a stored energy source (e.g. a battery bank and electric motor AC /DC).

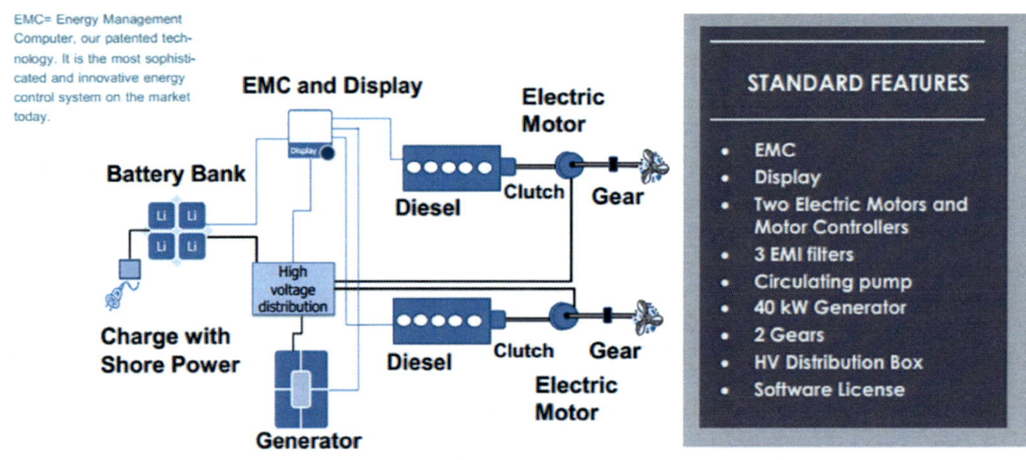

Hybrid Configurations

- There are three basic Hybrid configurations
    - Engine (Gas/Diesel)/Electric-AC
        » Engine – Electrical (Direct)
        » Engine – Electrical (Hybrid With Inverter and Batt Bank)
    - Engine (Gas/Diesel)/Electric-DC
    - Parallel Hybrid (Engine or Electric Motor)

Engine - Electrical (AC Motor)

The engine (e.g. petrol or diesel) is connected directly to an electrical generator. From this point on the power in the system is transferred electrically to the propeller shaft via a motor controller and electric (AC)motor

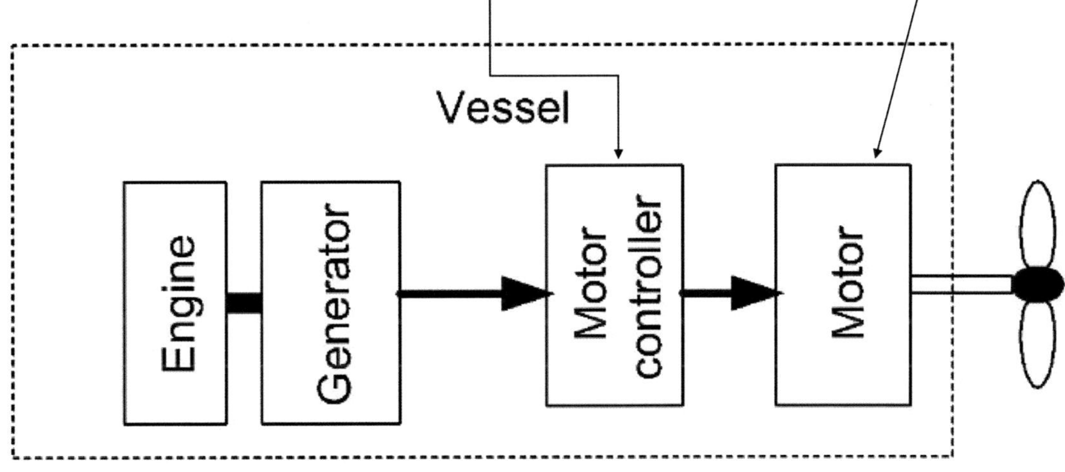

Engine-Electrical (AC Motor)

Strictly speaking this is not a hybrid due that there is no electric storage of energy The system may have multiple generators and multiple motors connected to a common electrical bus. This technology is used in diesel/electric trains and many large ships

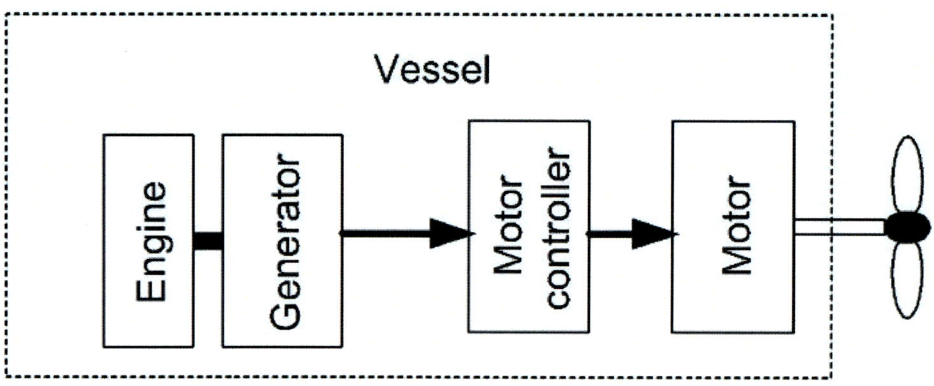

Engine-Electrical (AC Motor)

In this design , the boat propeller is moving only if a generator is running

The motor controller could be : Hydraulic, Mechanical or Electromechanical.

The shift control is connected directly to the motor controller

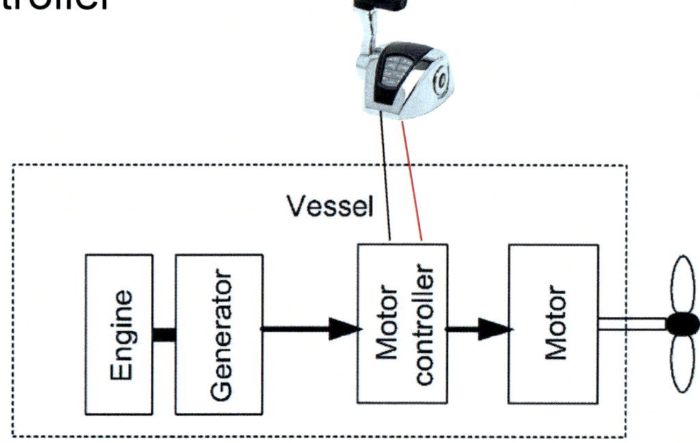

Hybrid Electrical (12 / 24 V DC)

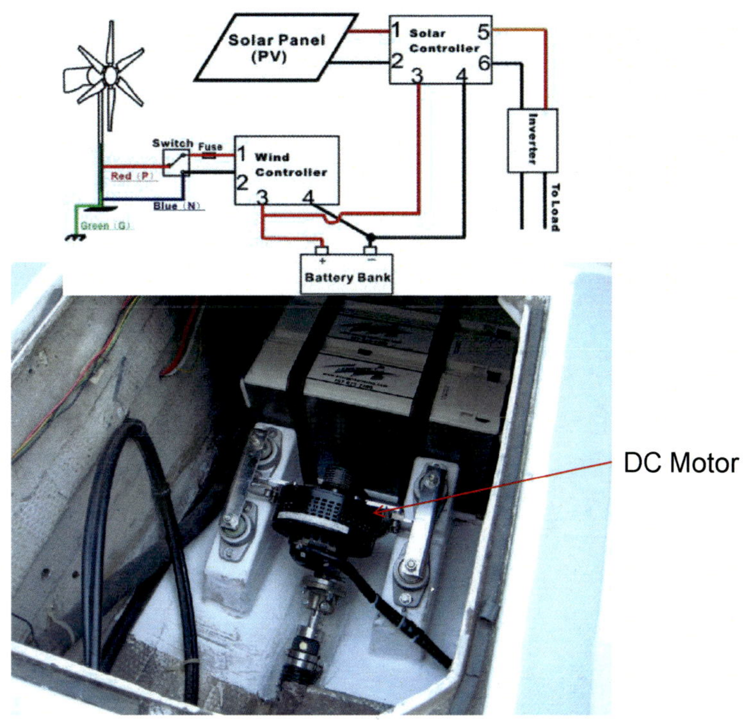

Hybrid Electrical

In this case a battery bank is also connected to the common electric power buss. In this system you can stop the engine and use the stored energy in the battery bank. With large batteries you can have long periods of electric propulsion (and/or driving onboard electrical appliances) without resorting to the generator.

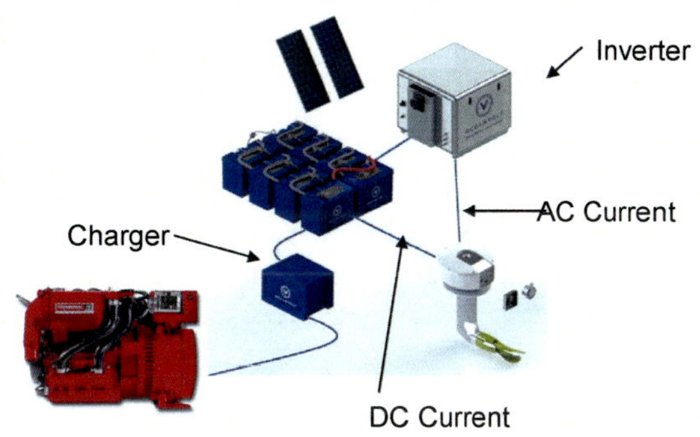

Hybrid Electrical

A dual function electric motor (AC/DC) is powered directly from the battery bank only when the generator is off and the battery bank charge is over 80%

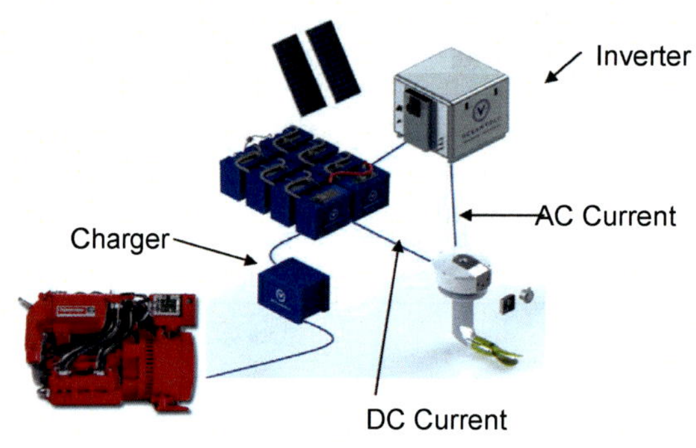

Fully Electrical

When the generator is running the boat works at the AC mode. The propulsion motor is powered from the inverter , the inverter from the battery bank and the battery bank from the generator

Depending of the battery bank capacity the system can operate without the generator for a limited period of time

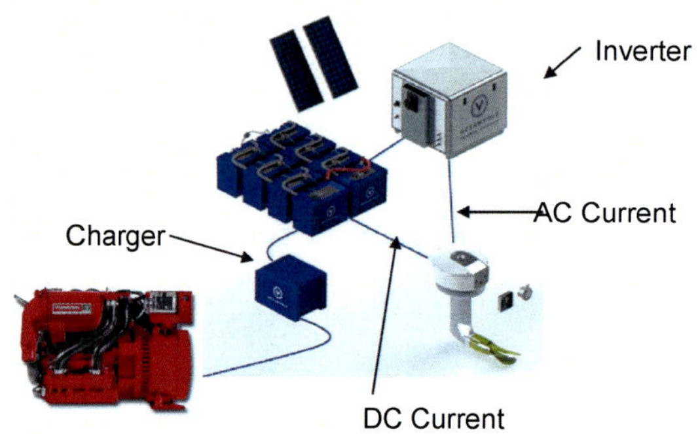

Hybrid Electrical

When the engine is running the electrical module (Blue section) works as a generator producing AC power for the home

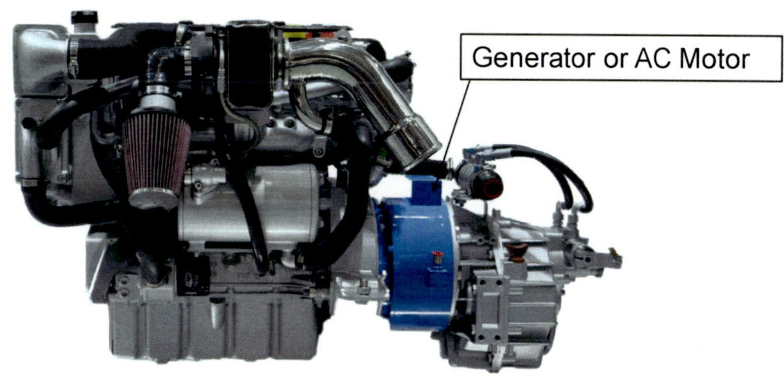

Hybrid Electrical

When the engine is off, then the blue section works as an electric AC motor powered from the generator or from the inverter

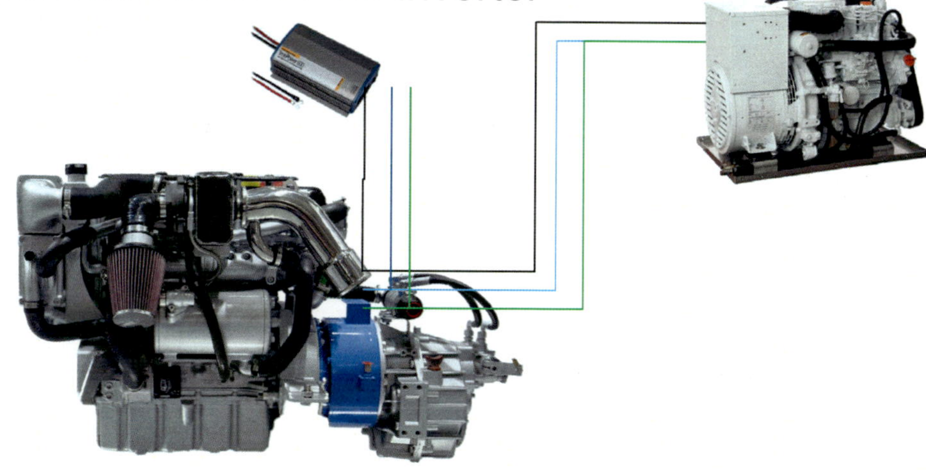

Hybrid Electrical (AC Motor With Inverter)

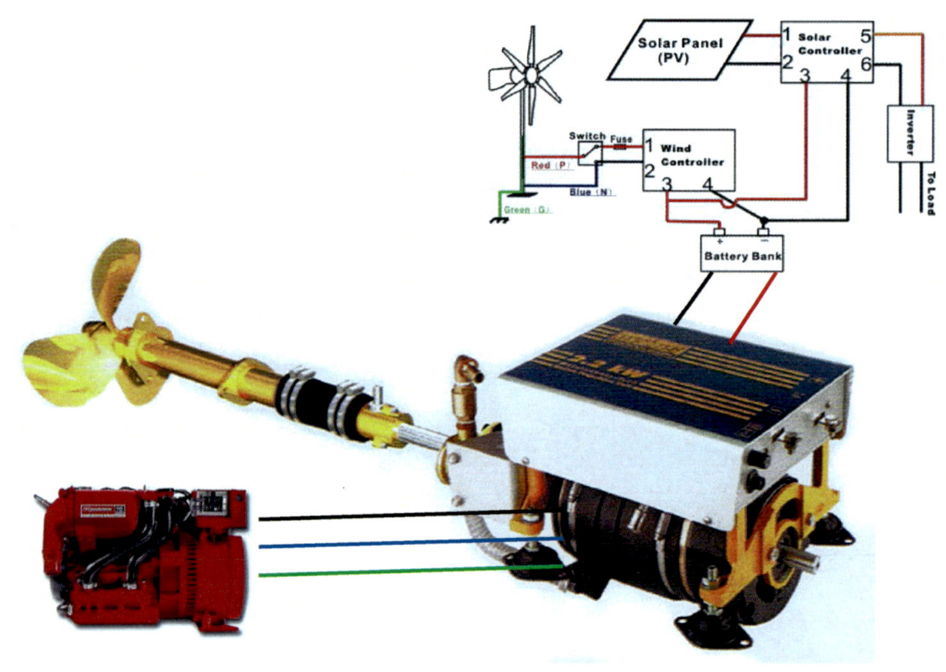

Hybrid (Hydraulic)

Main Engine
Or
Generator
Or
Electric Motor

Central Hydraulic Unit

Hydraulic Motor

Chapter 2
Marine Transmission Classification

Classification

- The marine transmission are classified according with their location in the boat as: In-board transmissions and Out-board transmissions or lower units

- Additionally the inboard transmissions possess different gear ratios and output torques depending if they are designed for diesel or gasoline engines

In-Board Transmissions

The inboard transmissions are classified according with the configuration between the motor and the propeller shaft into: a) In-line shaft , and b) V-drive shaft

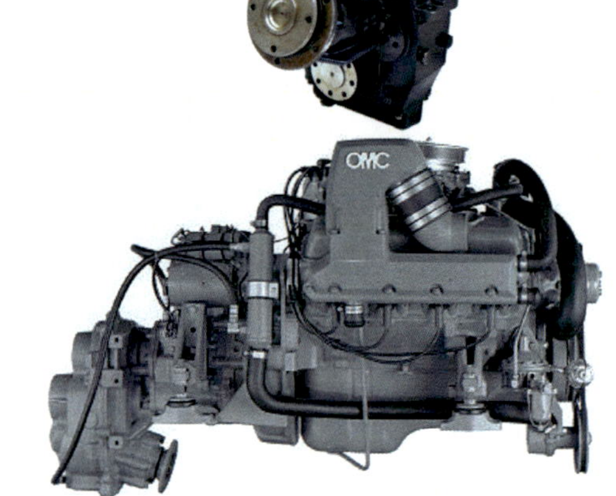

Gear Box or Transmission ?

- In technical terms the gear box is part of the transmission . A gear box is basically a synchronized group of toothed wheels that is used to transfer motion from one shaft to another . And a Transmission is a gear box controlled, manually , hydraulically or servo-assisted

- The marine transmission is a perfect combination of the three systems with a constant gear ratio on forward and reverse

In-board Electric Motor

The internal combustion engine can be replaced by an Electric motor, however the transmission is required for some reasons such as

- Increasing at the output torque
- Change in direction of rotation on the transmission instead of the engine
- Protection of the motor in case of propeller collision

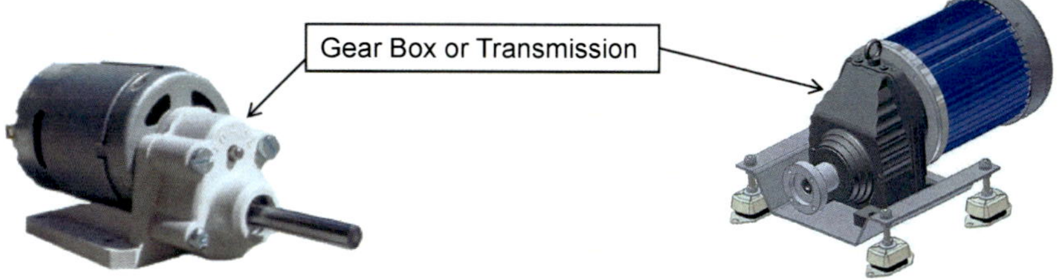

In-line Transmission

Components

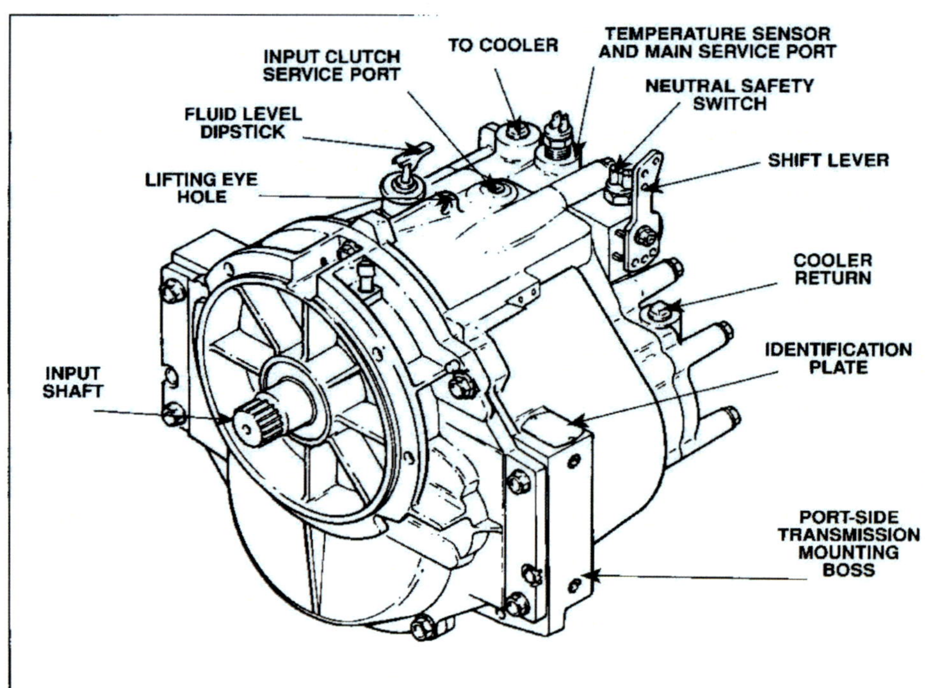

In-line Transmission (Parallel Axis)

In this configuration the input and output shafts are parallels , with a short separation due to the gears and bearings thickness

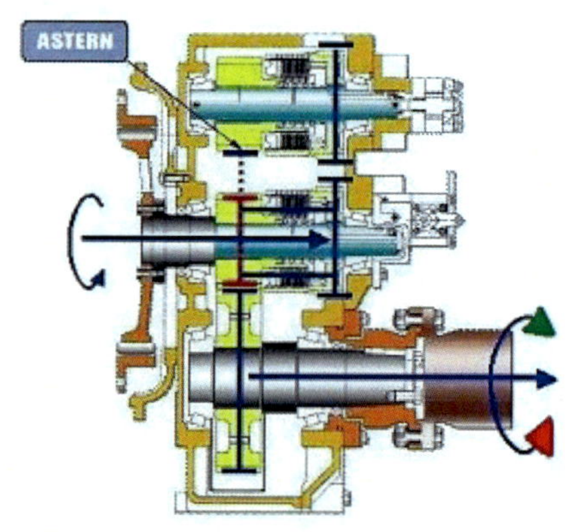

In Line Transmission (Straight axis)

In this configuration the input and output shafts are in the same axis due to the use of hollow shafts with plain bearings

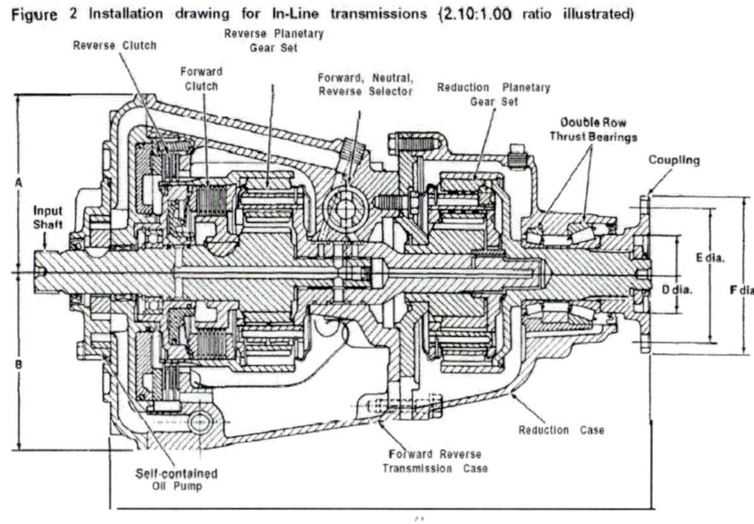

In Line Transmission (Straight axis)

This types of transmission are used in engine rooms where the separation between the ceiling and the bilge is reduced

In Line Marine Transmission

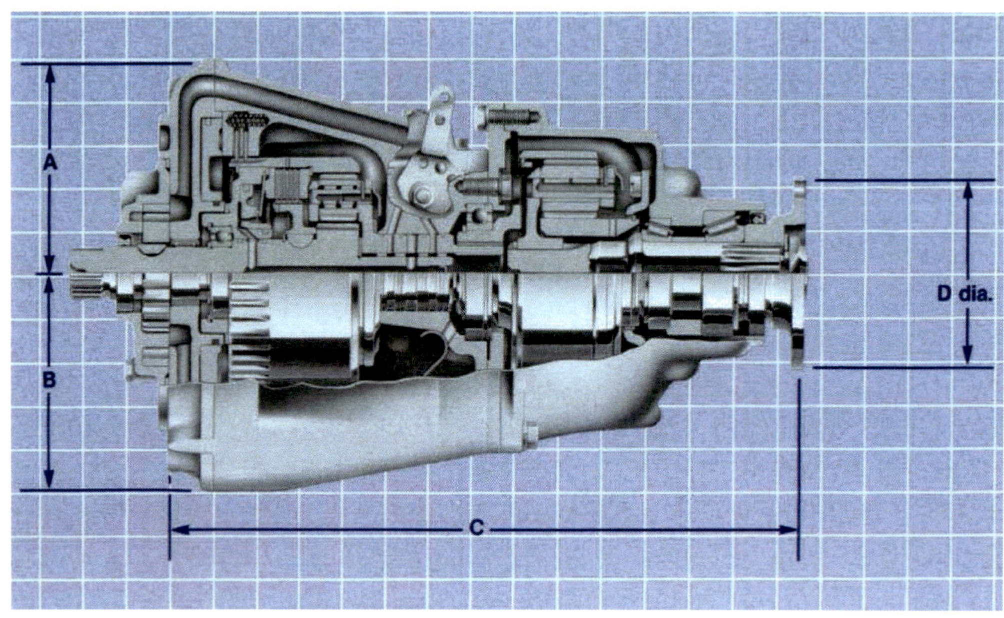

In-Line Shaft Configuration

The engine and transmission/drive unit are inside the boat. A drive shaft is connected at the rear of the transmission and is run out of the hull through a sealing unit. The propeller is connected directly to the drive shaft at the in-water end. A separate rudder steers the boat

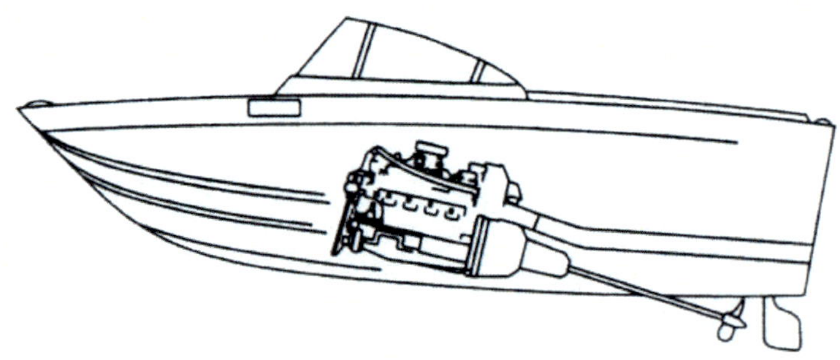

In-Line Shaft Configuration

Compared with all the shafts, U-joints, bellows and steering components of a sterndrive, an inboard (even one with a V-drive transmission) is less complex. There are fewer elements to break, wear out or service. Also, corrosion is less of a concern with inboard running gear than with an aluminum sterndrive.

In-Line Transmission

The in-line transmission is gear set, driven by a complex hydraulic circuit composed by : clutch packs, servo valves, planetary gears, actuators, solenoids and hydraulic pumps

Disadvantages

An in-line running gear system generates a lot of drag, reducing speed and miles per gallon versus a sterndrive

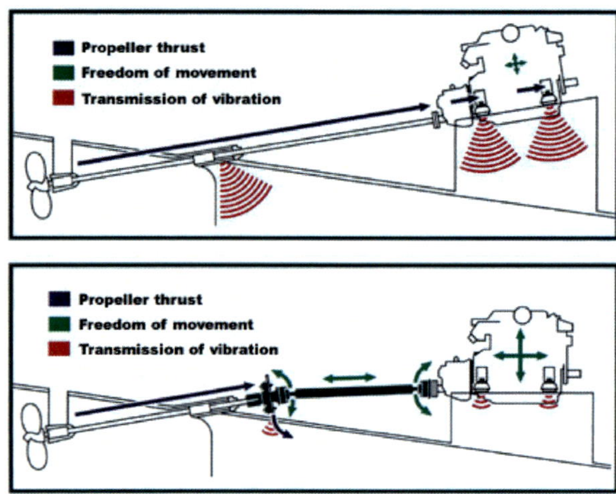

V-Drive Transmission

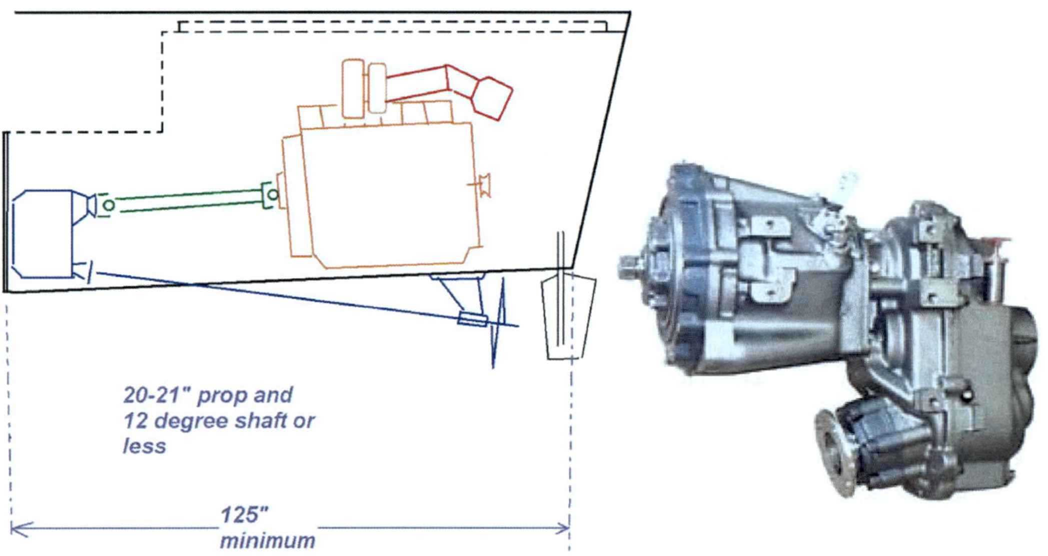

V-Drive Transmission

Advantages:

- Greater safety (with propeller tucked under the boat, not under swim platform)
- Improved handling, reduced drag and bow rise
- Shallower draft
- Increased cockpit space inside the boat
- Reduced maintenance and increased reliability compared with sterndrives

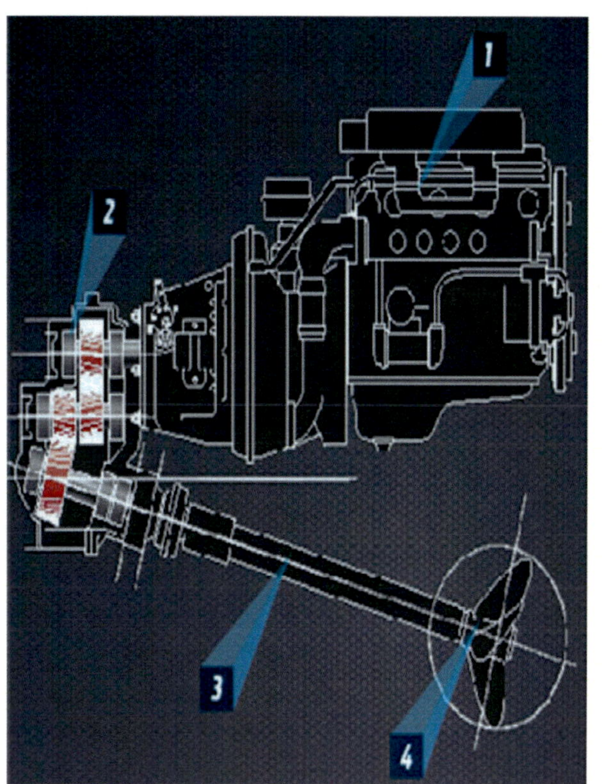

V-Drive Transmission

- Because the V-Drive requires no universal joints, a potentially costly maintenance item is completely eliminated

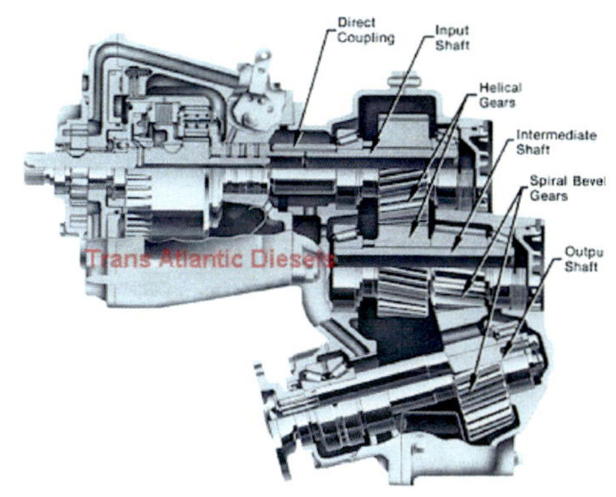

V-Drive Transmission

A series of reduction ratios is available. For twin-screw applications, opposite propeller rotation can be provided for two engines of like rotation through the use of gear and chain drive transmissions.

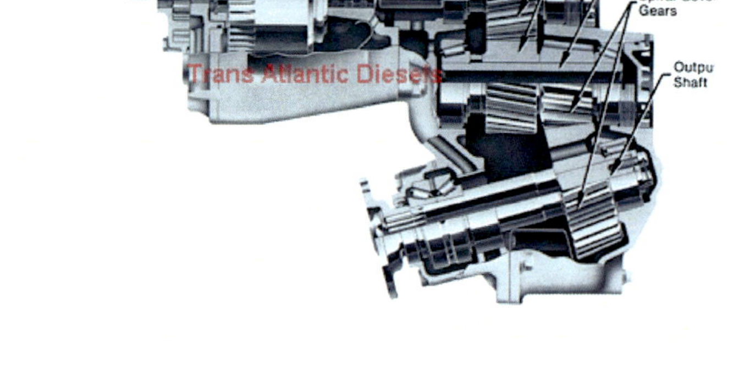

V-Drive Configuration

The V-drive transmission systems of yesteryear possessed a bad reputation for their racket and gear noise. However, today's V-drive transmissions from companies such as Twin Disc, Velvet Drive and ZF offer smoother and quieter operation. Gearing can be specified to suit the boat, engine and style of boating

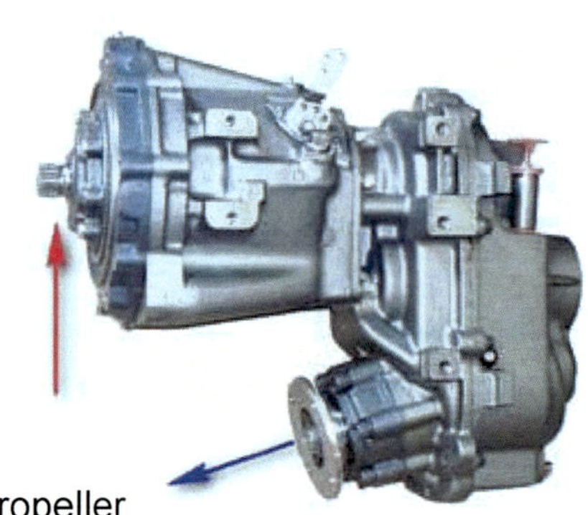

V-Drive Configuration

The engine is mounted with the front (pulley end) facing stern (rear). The transmission faces the bow (front) and by the use of a universal joint on the end of the transmission, the drive shaft is directed sternwards under the engine and through the hull. The propeller/rudder arrangement is the same as in a standard inboard installation

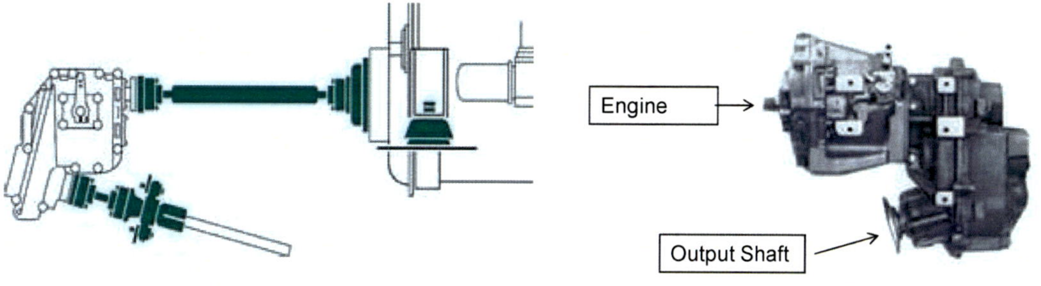

V-Drive Vs Stern-Drive

Why?. because an inboard running gear generates a lot of drag, reducing speed and miles per gallon versus a sterndrive. Also, a sterndrive's angle of thrust is virtually horizontal, further increasing efficiency over the 7- to 12-degree down angle of an inboard's prop shaft.

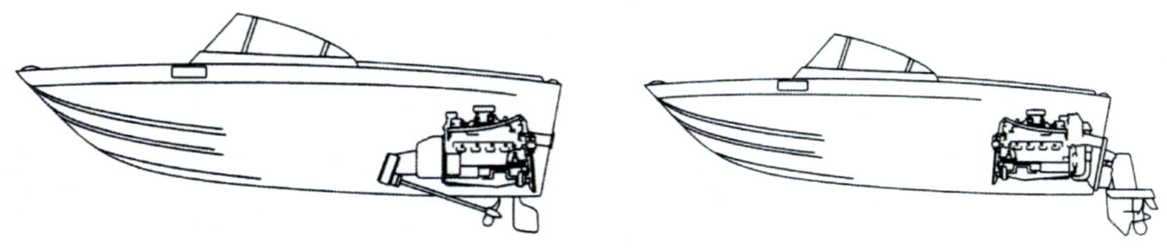

Inboard/Outboard Drive System

Stern drives are used for small to large ski boats, large cruisers, high performance, and heavy boats needing a heavy duty drive

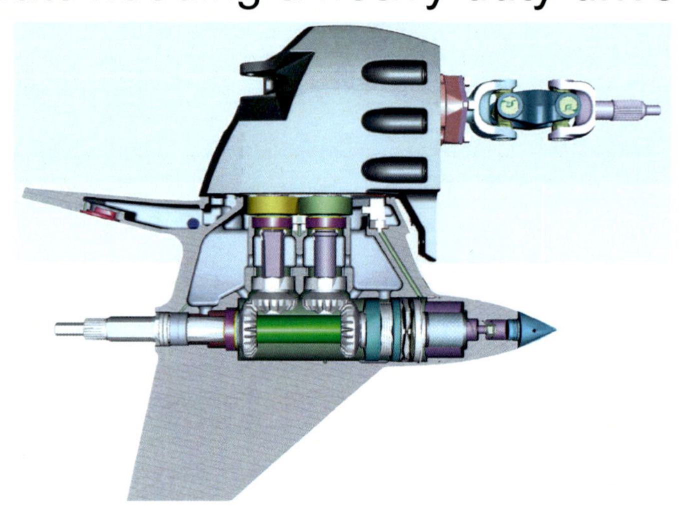

Stern-Drive Systems

Stern drives are available in a few different options to be used for small light weight, lower horsepower boats, all the way through heavy high horsepower boats with Diesel Engines

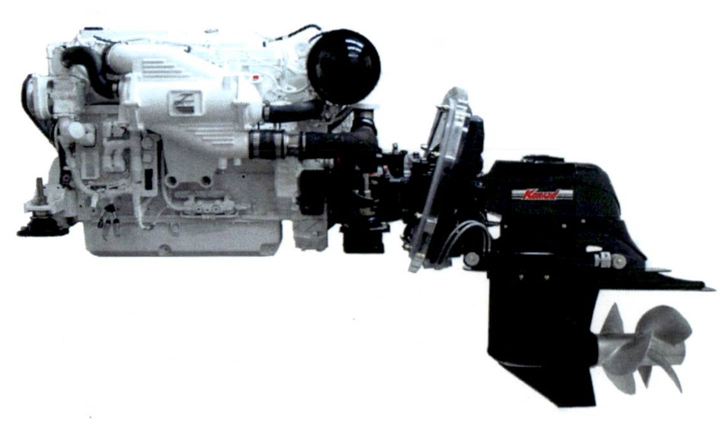

Stern-Drive

The engine is mounted inside the boat. The transmission/drive unit is outside, attached to the transom. The engine and drive unit are joined by a torque coupler and universal joint. Angling the drive steers the boat

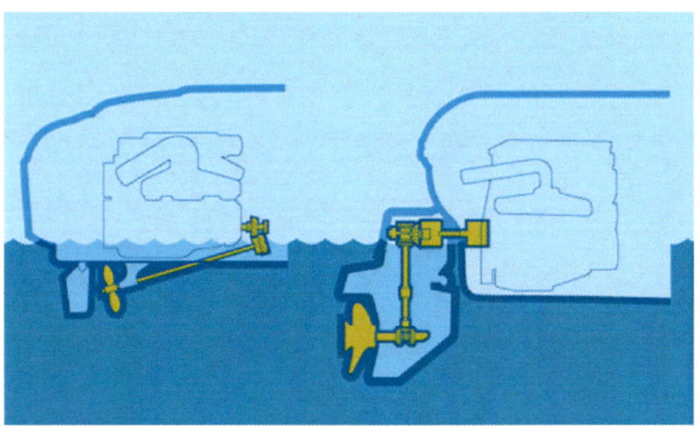

V-Drive Vs Stern-Drive

A V-drive is more efficient in delivering horsepower to the prop, losing only 8 percent, while a stern-drive loses 13 percent. Yet, all other things being equal, a stern-drive is faster and more fuel-efficient

V-Drive Vs Stern-Drive

While a sterndrive offers better performance, the anti-ventilation plate limits its propeller diameter. With a V-drive, you're not as limited, and this can be an advantage if, for example, you want to bump up the prop diameter to help improve low-speed handling

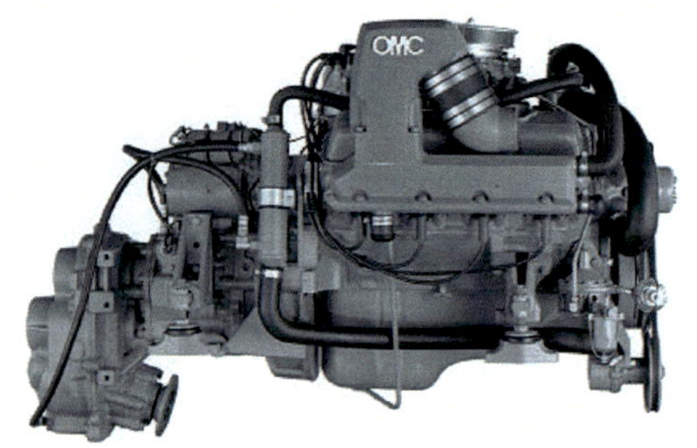

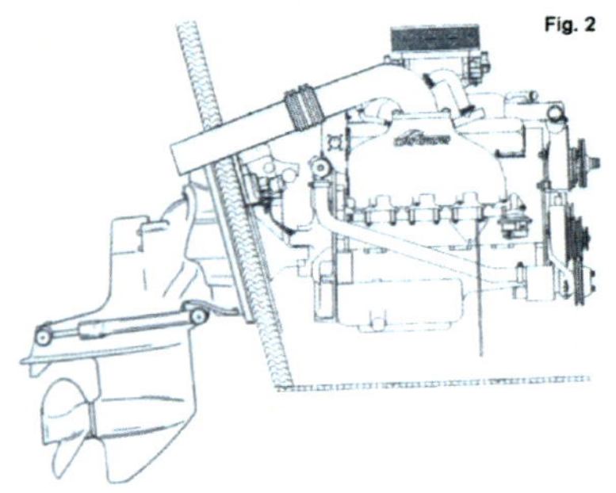

In-Line Drive Vs V-Drive Vs Stern-Drive

V-drive systems offer more space amidships than a conventional inboard and require less maintenance and service than sterndrive systems. While sterndrives are faster and more fuel-efficient, a V-drive can swing a bigger wheel, and you're more likely to find a replacement prop when far from home

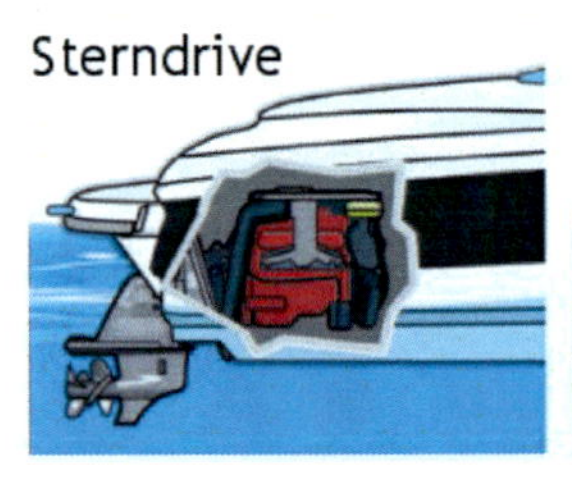

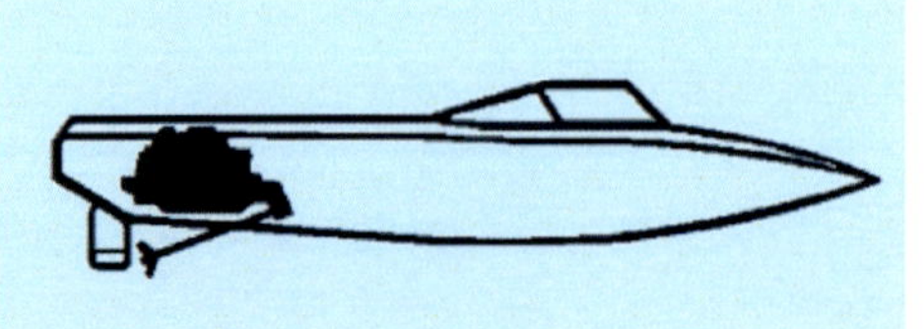

Inboard Performance System

Advantages

- Improved efficiency, higher top speed, reduced fuel consumption/extended range, and great acceleration
- Substantially enhanced maneuverability regardless of speed and excellent course keeping capability offers safe and predictable handling
- Much improved overall environmental care
- Onboard comfort is greatly enhanced thanks to much lower levels of sound, vibrations and exhaust fumes

In board Performance System (IPS)

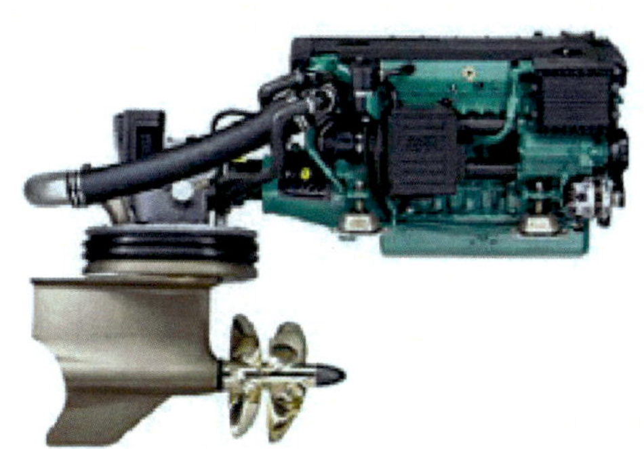

Advantages

- Installation is greatly simplified
- More space available for cargo
- Improved safety and redundancy
- Ease of service, the complete system from helm to propeller cone supported by one supplier

In board Performance System (IPS)

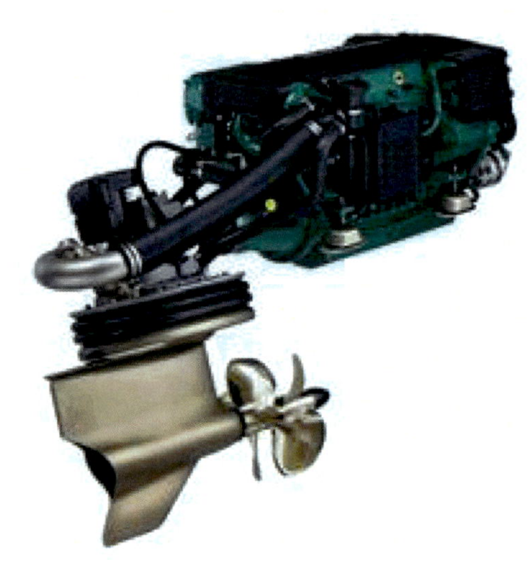

Volvo Penta IPS 500

The propellers are forward facing at the front of a hydro-dynamically optimized pod unit, working in undisturbed water, with a minimum of pressure pulses affecting the hull

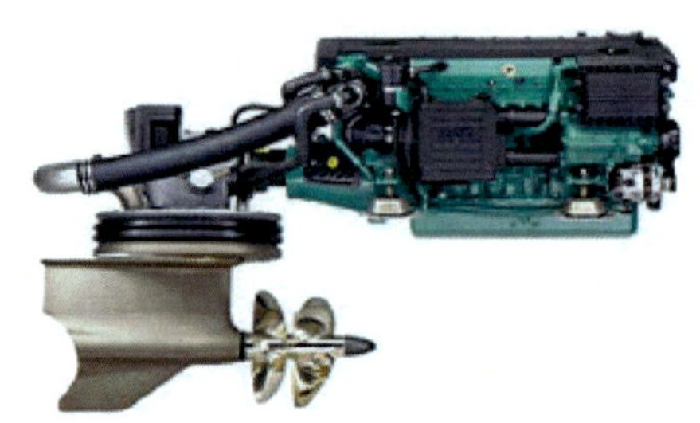

Volvo Penta IPS 500

The pod unit is designed for an efficient flat hull surface with maximum buoyancy. The propeller thrust is parallel with hull and all the power is driving the boat forward

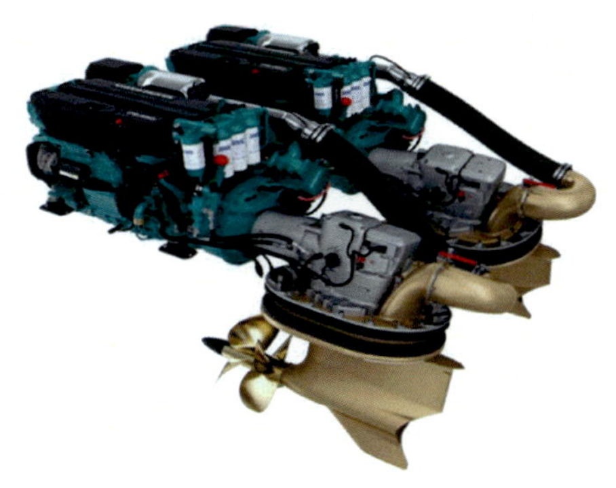

Volvo Penta IPS 500

The propeller position is well under the hull giving minimum risk for cavitation caused by of air intrusion and less marine growth

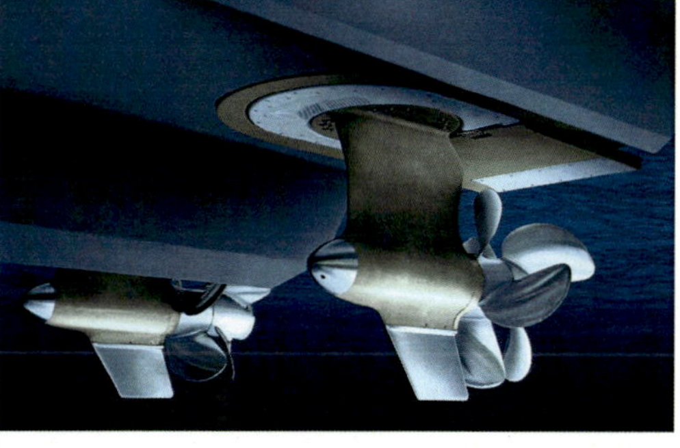

ZF POD - Pod Drive System

- The ZF POD is one of the most efficient propulsion system, ideal for medium-sized pleasure craft of 30 to 48 feet overall length, can be used with Diesel engines up to 450 hp

Electrical (IPS)

Some electrical designs isolate the transmission from the electrical motor by means of a hydraulic or electric clutch pack

Jet-Drive

The engine is mounted inside the boat and attached to a high speed pump which draws in water through an intake grate, increases its velocity and forces it out through a directionally adjustable nozzle which is mounted outside the boat. Changing the direction of the nozzle steers the boat

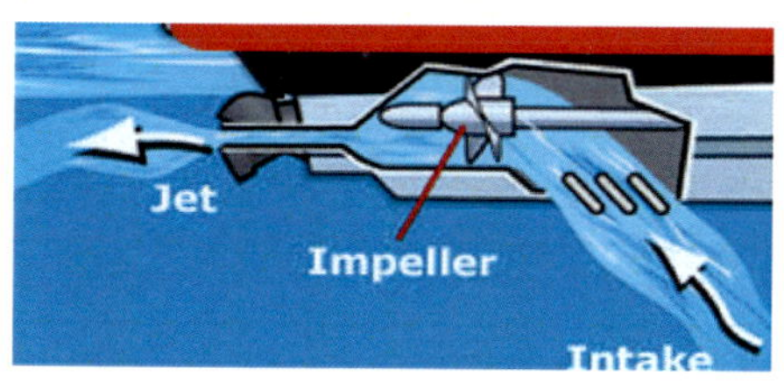

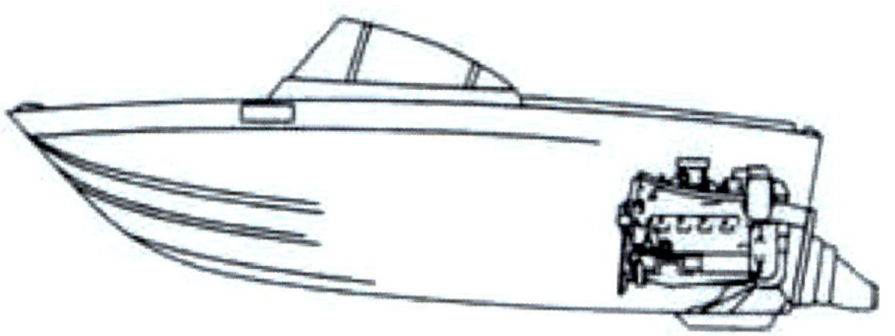

Jet Drive

A jetboat draws the water from under the boat through an intake and into a pump-jet / Impeller-jet inside the boat, before expelling it through a nozzle at the stern

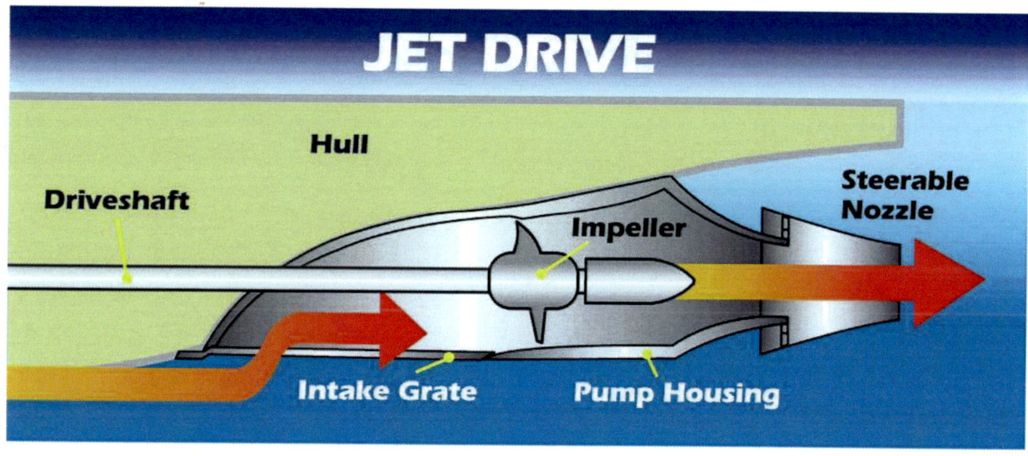

Jet Drive

a waterjet unit delivers a high-pressure "push" from the stern of a vessel by accelerating a volume of water as it passes through a specialised pump mounted above the waterline inside the boat hull

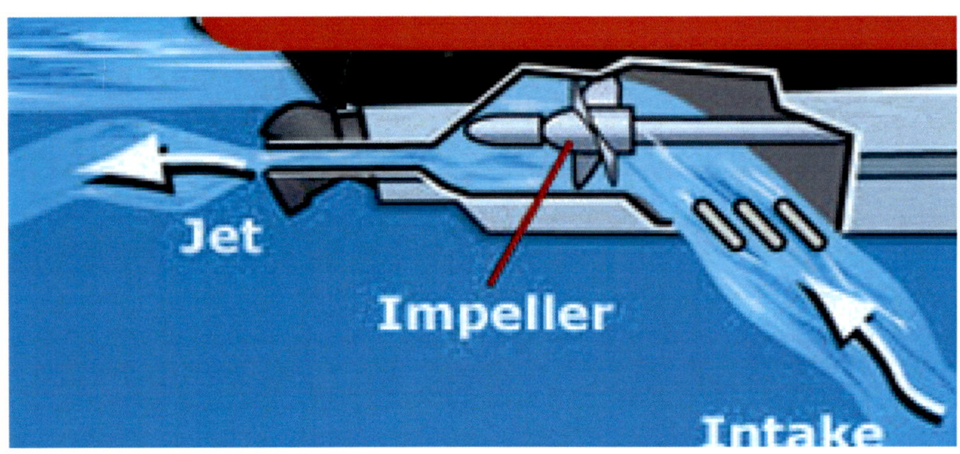

Jet Drive

In a jetboat, the waterjet draws water from beneath the hull, where it passes through a series of impellers and stators - known as stages - which increase the velocity of the waterflow

Jet Drive

The jetstream exits the unit through a small nozzle at high velocity to push the boat forward. Steering is accomplished by moving this nozzle to either side, or less commonly, by small gates on either side that deflect the jetstream

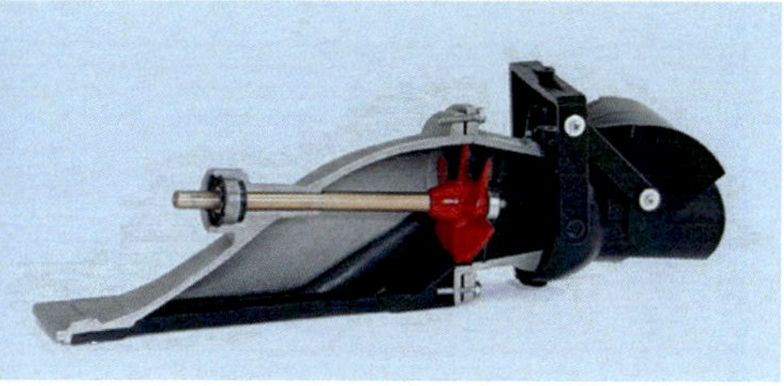

Jet Drive

This deflector redirects thrust forces forward to provide reverse thrust. Most highly developed reverse deflectors redirect the jetstream down and to each side to prevent recirculation of the water through the jet again

Chapter 3
Manual Transmission

Transmissions

- In order to understand how the inboard marine transmission works we are going to study the automotive manual transmissions and the automotive automatic transmissions with their respective clutch systems

- The marine transmissions combine both types of systems . However there is still a long way to go in terms of research in these mechanical transfer systems

Manual & Automatic

Both the **Automatic transmission** and a **Manual transmission** accomplish exactly the same thing, but they do it in totally different ways. It turns out that the way an automatic transmission does it is absolutely amazing!

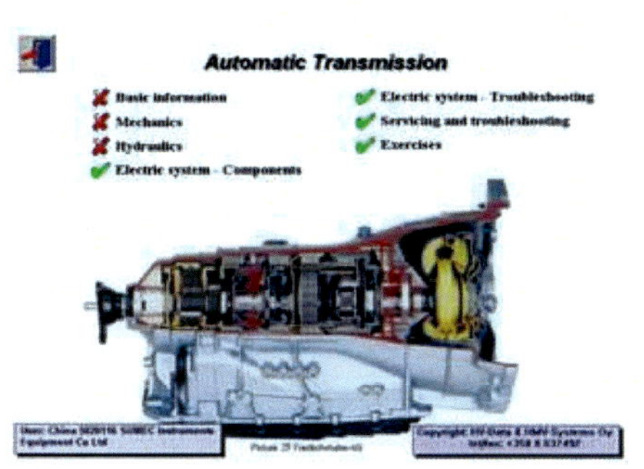

Manual Transmission

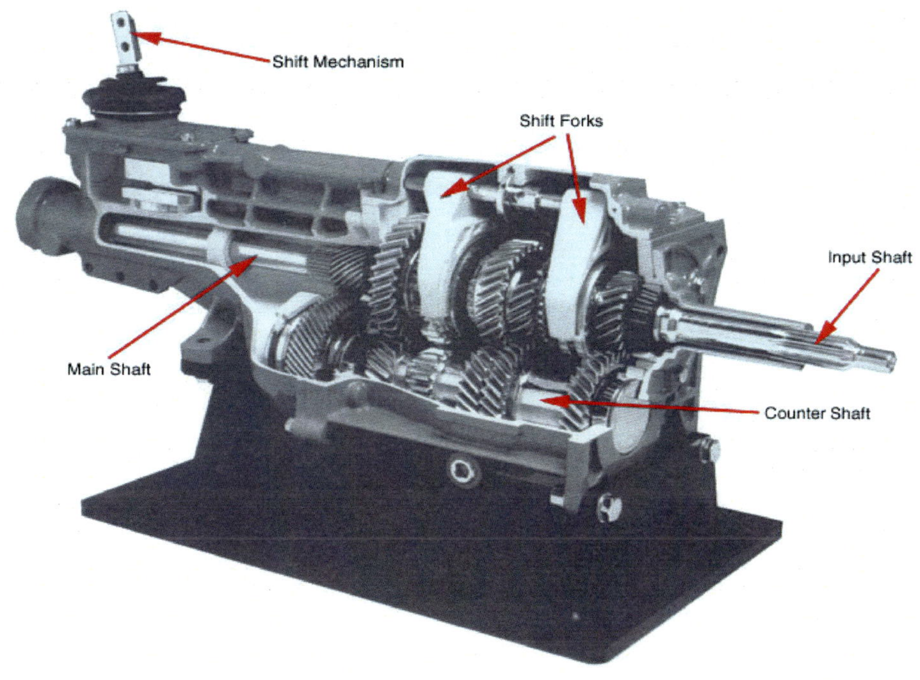

Automotive Transmission

The transmission allows the gear ratio between the engine and the drive wheels to change as the car speeds up and slows down

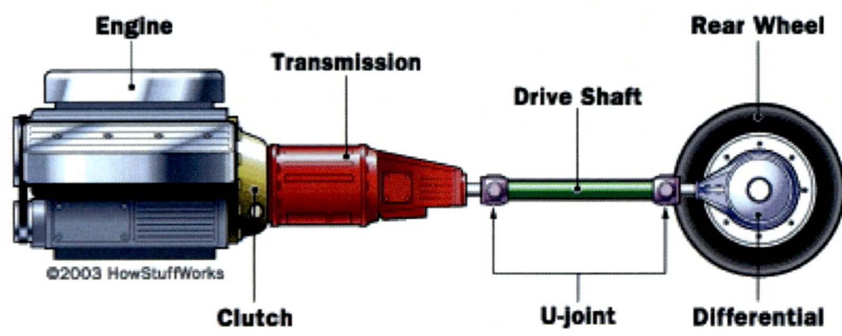

Marine In-board Transmission

The transmission allows the gear ratio between the engine and the drive wheels to change as the car speeds up and slows down

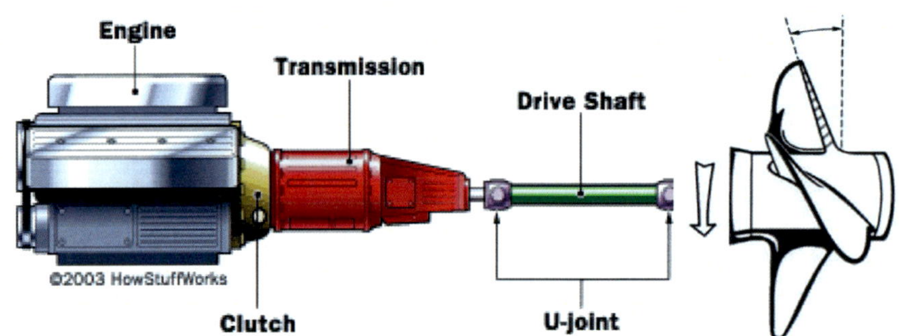

Mechanical Clutch

- Clutch Parts and Assembly
- Types of clutches
- How the Clutch works
- The Pilot Bearing
- The Pressure Plate
- The Driven Plate
- The Friction Disc
- The Clutch Release Bearing
- Mechanical and Hydraulic Clutch Pedal

Mechanical or Diaphragm Clutch

In everyday use, the term clutch refers to a subcomponent of motor vehicle engine's transmission designed to allow engagement or disengagement of the engine to the gearbox or whatever apparatus is being driven

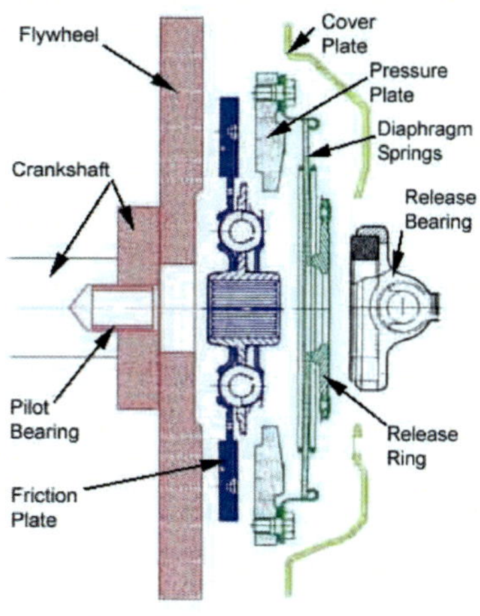

The Clutch

The clutch is located between the engine and the gearbox, as disengaging it is required to change gear. Although the gearbox does not stop rotating during a gear change, there is no torque transmitted through it, thus less friction between gears and their engagement dogs

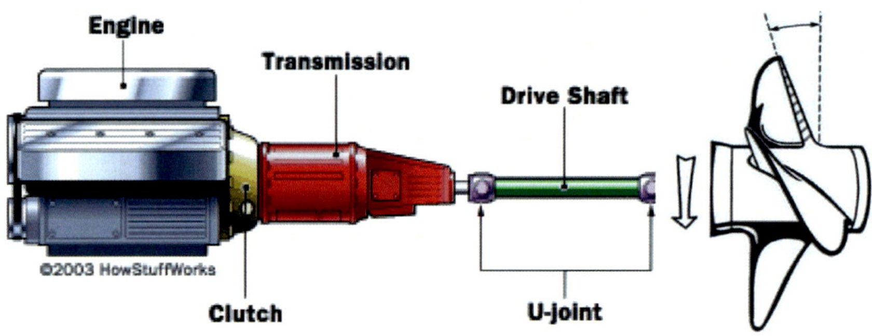

Clutch Assembly

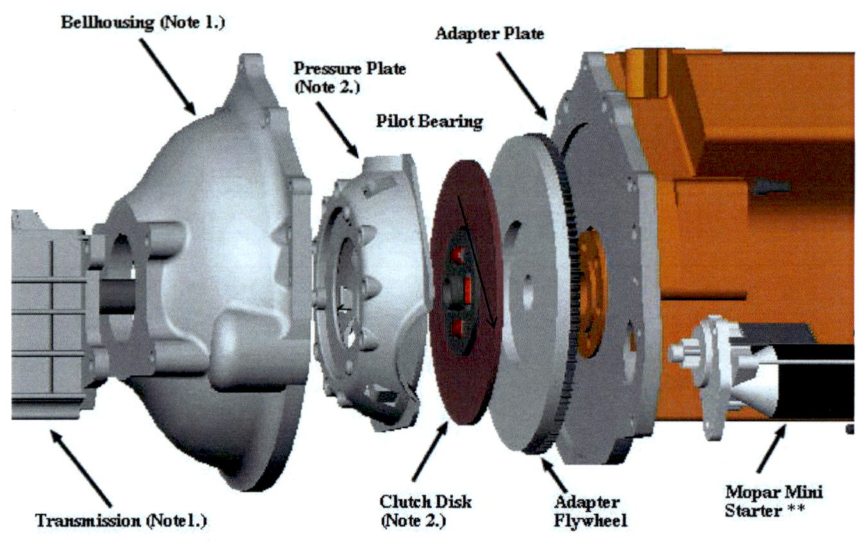

The Clutch Parts

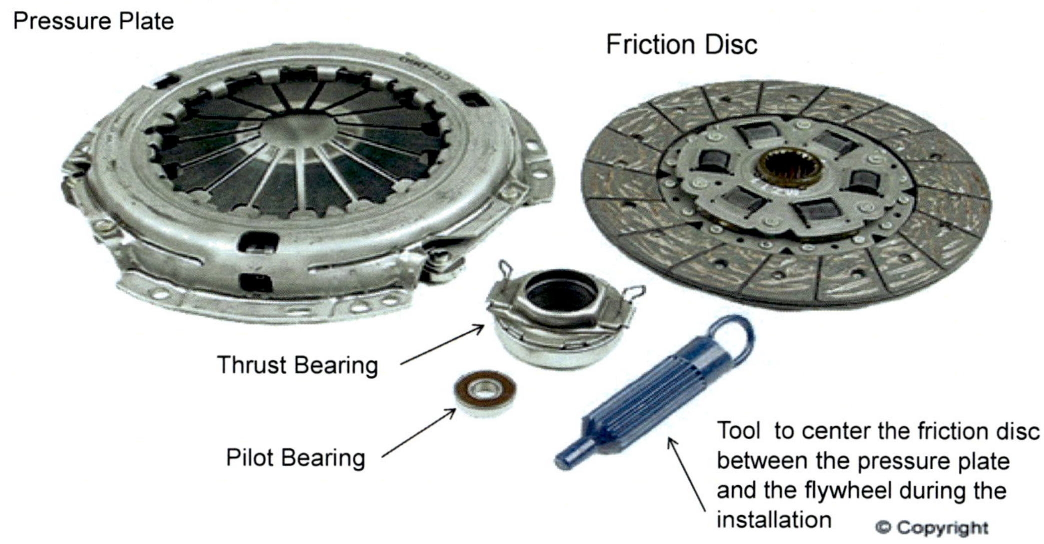

Types of Clutches

- A **wet clutch** is immersed in a cooling lubricating fluid, which also keeps the surfaces clean and gives improved performance and longer life.

- A **dry clutch**, as the name implies, uses no fluid.

- No pressure on the pedal means that the clutch plates are engaged (driving), while depressing the pedal will disengage the clutch plates, allowing the driver to shift gears.

Types of Clutches

Mechanical clutches provide either a positive (no-slip) or a friction-dependent drive; **centrifugal** clutches provide automatic engagement and **Electromagnetic** clutches also provide automatic engagement

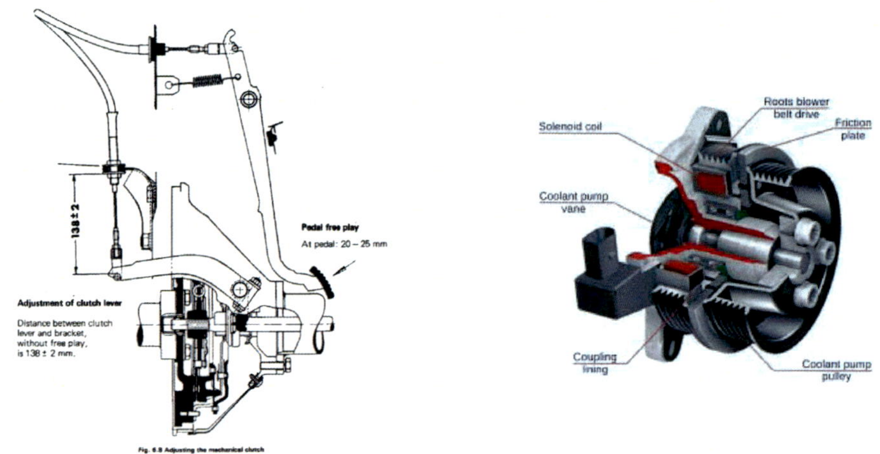

Clutches

A **clutch** is a mechanism for transmitting rotation, which can be engaged and disengaged.

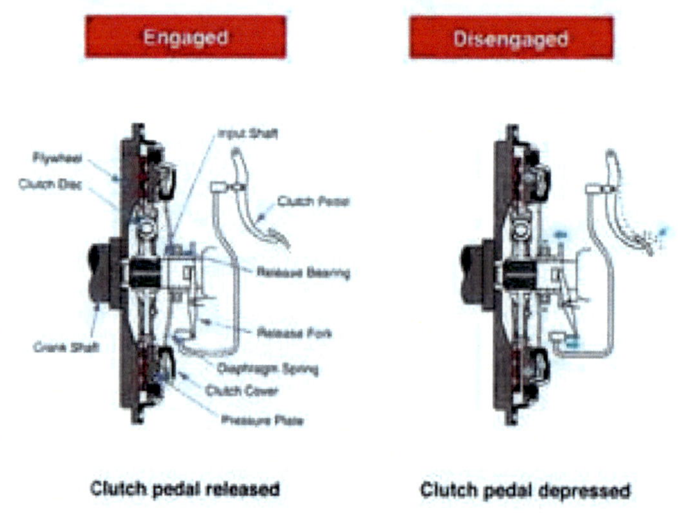

Clutches

Clutches in typical cars are mounted directly to the face of the engine's <u>flywheel</u>, as this already provides a convenient large diameter steel disk that can act as one driving plate of the clutch

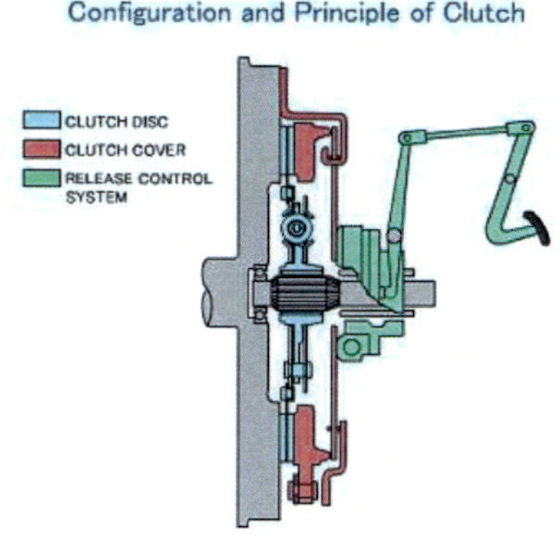

How the Clutch Works ?

When the clutch is applied , the pressure plate (1) force the friction plate (2) against the flywheel. As the friction plate works together with the input shaft of the transmission then from this moment the transmission is moving at the same rpm of the engine

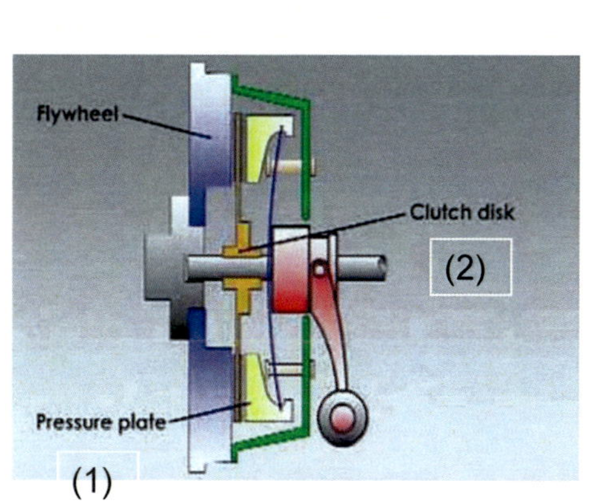

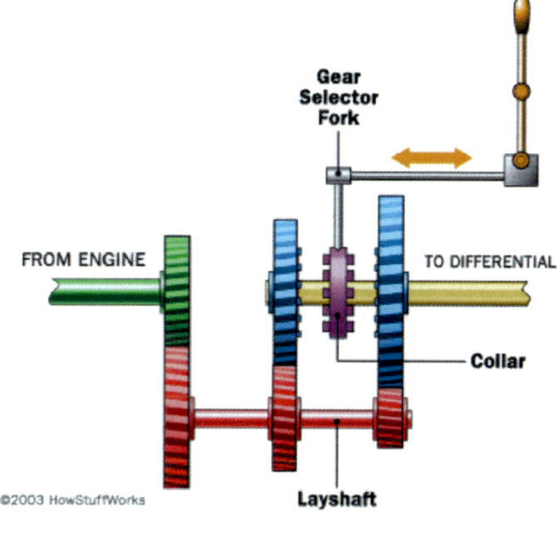

How the Clutch Works ?

Once again, when the pressure plate is released , then the transmission is isolated from the engine and you can select a new gear ratio

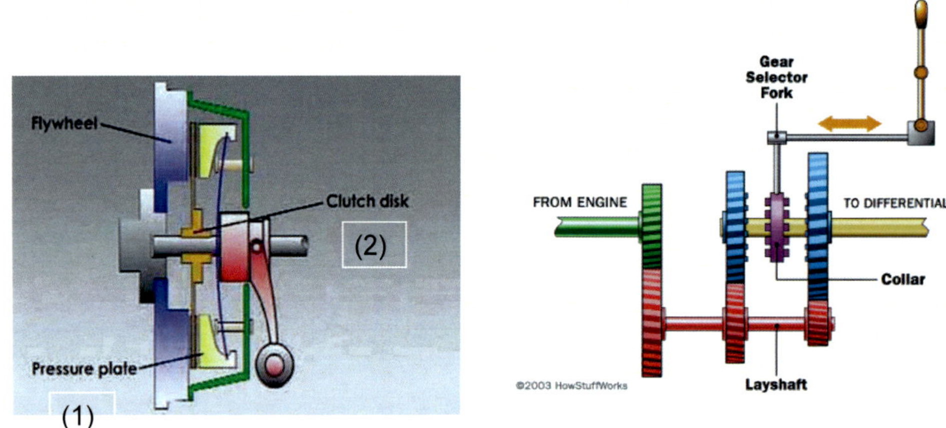

How Clutches works

The spring pressure is released when the clutch pedal is depressed thus either pushing or pulling the diaphragm of the pressure plate, depending on type, and the friction plate is released and allowed to rotate freely

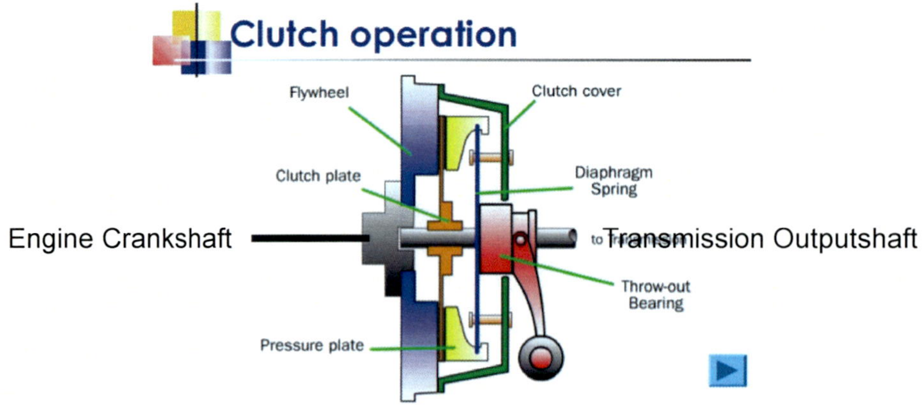

The Clutch Parts

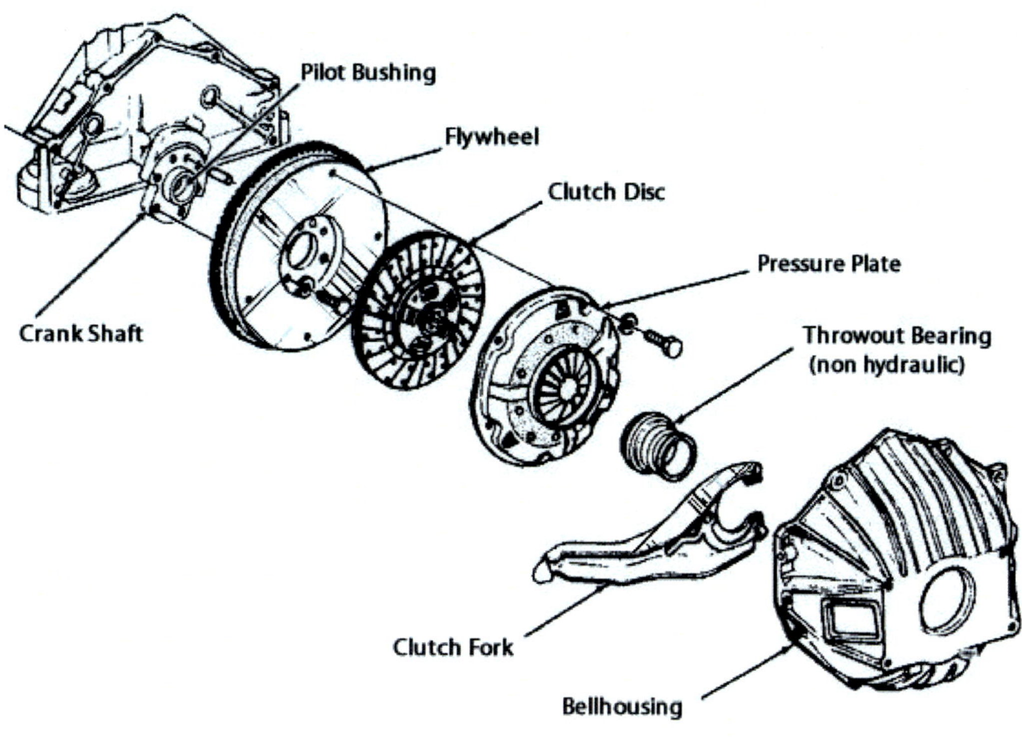

Pilot Bearing

The main role of a pilot bearing (or a pilot bushing) is to support the transmission input shaft as well as to align the clutch disc to the flywheel

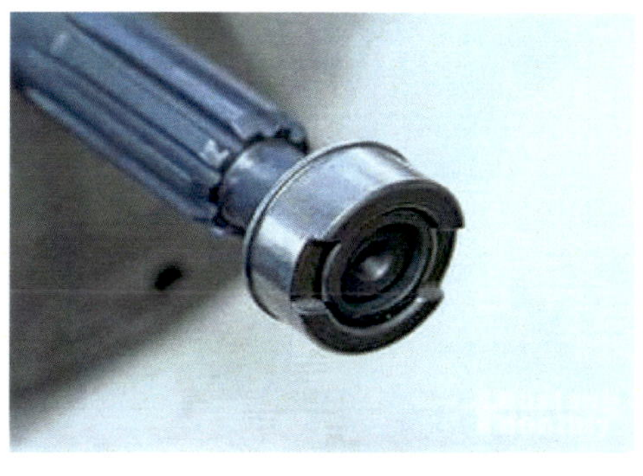

Pilot Bearing

A pilot bearing is mounted on the flywheel to support the input shaft of the transmission

Pilot Bearing

The input shaft of the transmission never is engaged   directly with the crankshaft. By means of the pilot bearing , the input shaft of the transmission spinning freely

Types of Pilot Bearings

There are three basic types of pilot bearings: needle bearing, roller ball bearing, and brass bushing

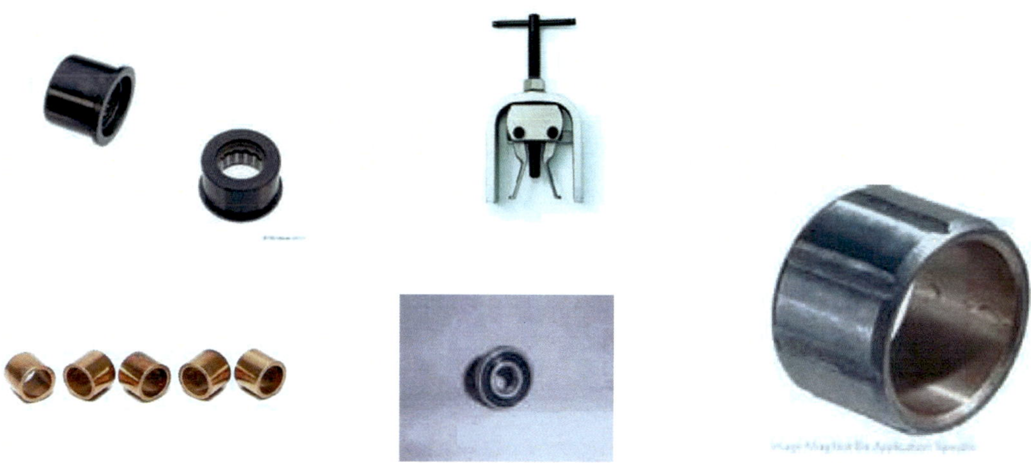

Pressure Plate

The pressure plate is a mechanism that serves to connect the friction plate with the flywheel. The friction plate is spinning freely between the pressure plate and the flywheel when the clutch is depressed

The Pressure Plate

The pressure plate (3) is bolted on the flywheel (1) . It is spinning constantly at the same speed with the engine. On the other hand the friction plate (2) works together with the input shaft of the transmission and it is spinning freely while the pressure plate remains disengaged

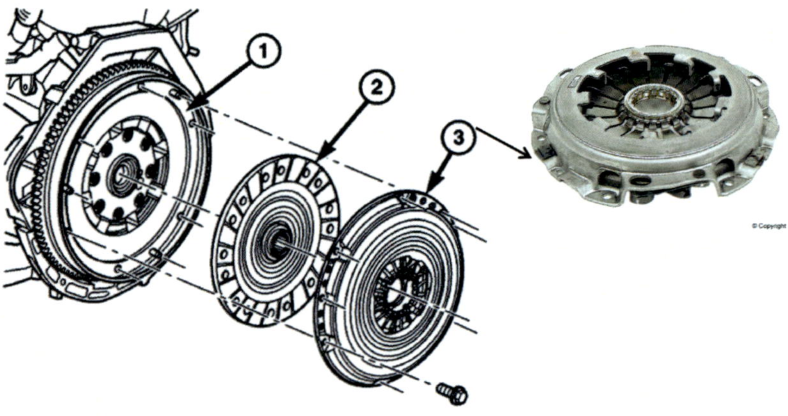

Pressure Plate

With the clutch disengaged, the gearbox input shaft is free to change its speed as the internal ratio is changed. Any resulting difference in speed between the engine and gearbox is evened out as the clutch slips slightly during re-engagement

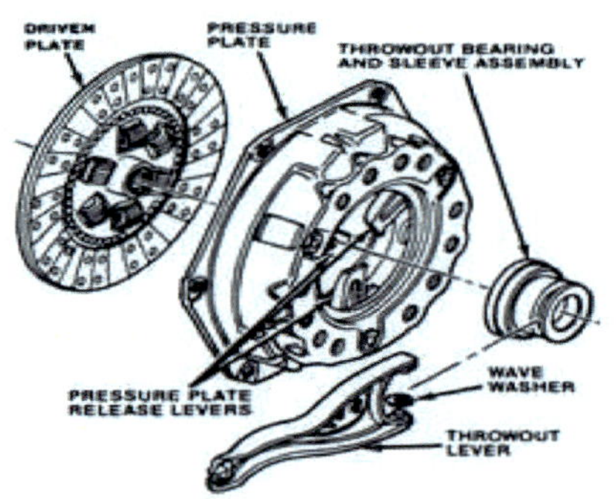

Driven Plate or Friction Plate

The driven plate is a friction type disc with two friction facings carried on waved spring steel segments and riveted to a steel disc. Drive is transmitted from the disc to a central splined hub, through a number of torsional coil springs.

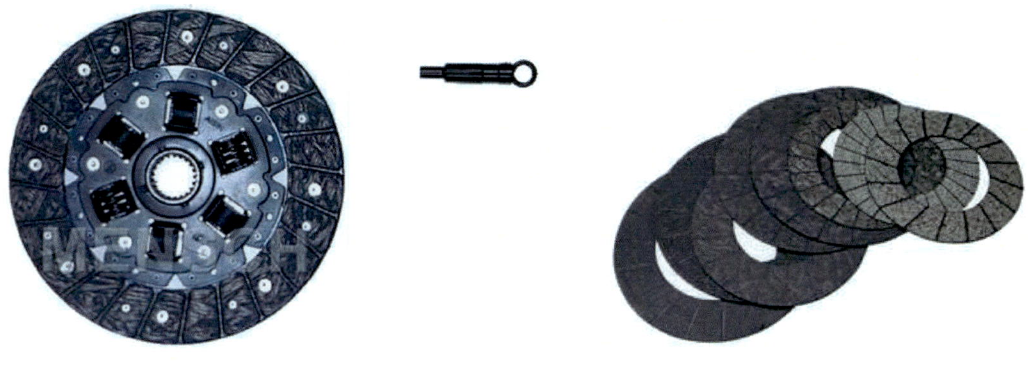

Friction Disc

The friction material is very similar to the material used in brake shoes and pads and used to contain asbestos and ceramics

The Friction Disc Functions

The friction disc has two functions: Engage the input shaft of the transmission with the crankshaft of the engine by means of pressure plate and absorb excessive vibration by means of the <u>torsional springs</u>

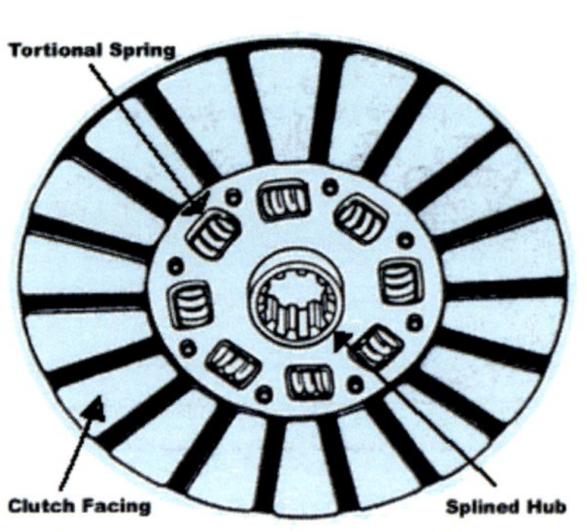

The Friction Disc Functions

The splined hub of the friction disc allows that the input shaft of the transmission and the friction disc be engaged at the same speed always

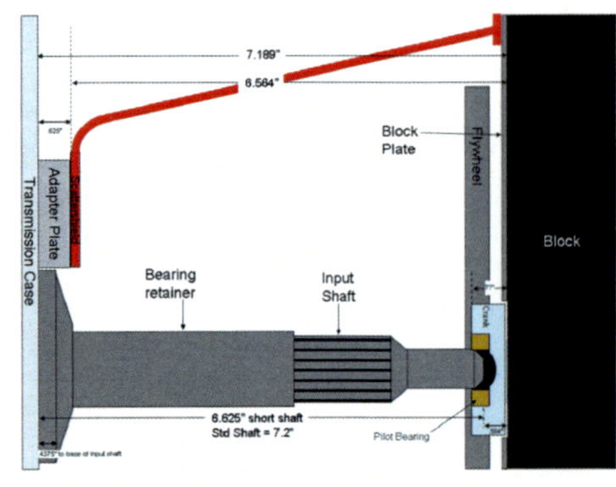

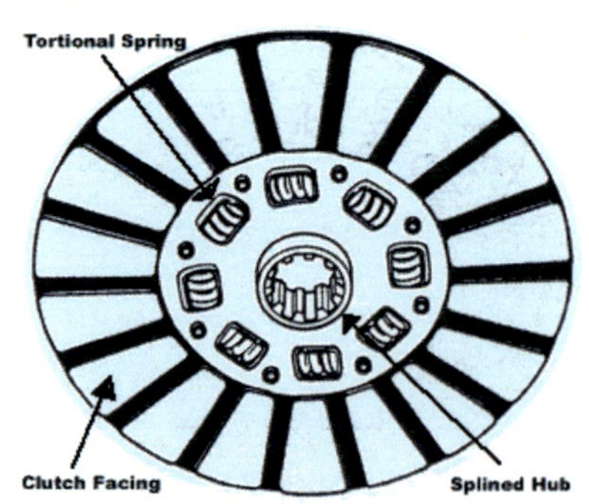

Input Shaft / Pilot Bearing

Finally , the input shaft of the transmission , spins freely on the pilot bearing. When the transmission is in neutral, the flywheel is spinning while the shaft remains free, allowing the driver to select the next gear position

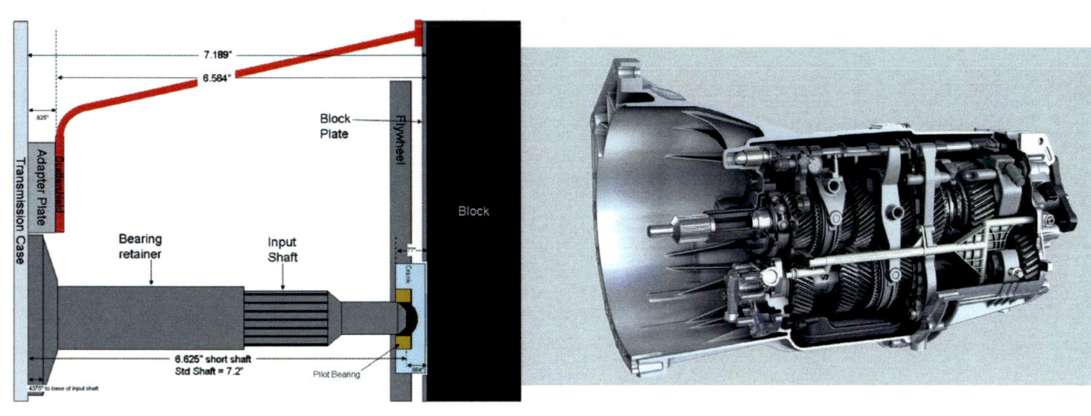

The Clutch Release Bearing

The clutch release bearing is an axial bearing capable to apply pressure on the diaphragm of the pressure plate in order to engage or disengage the friction plate

The Clutch Release Bearing

The bearing slides freely back and forward through the input shaft of the transmission by means of the fork that is moved by the clutch pedal

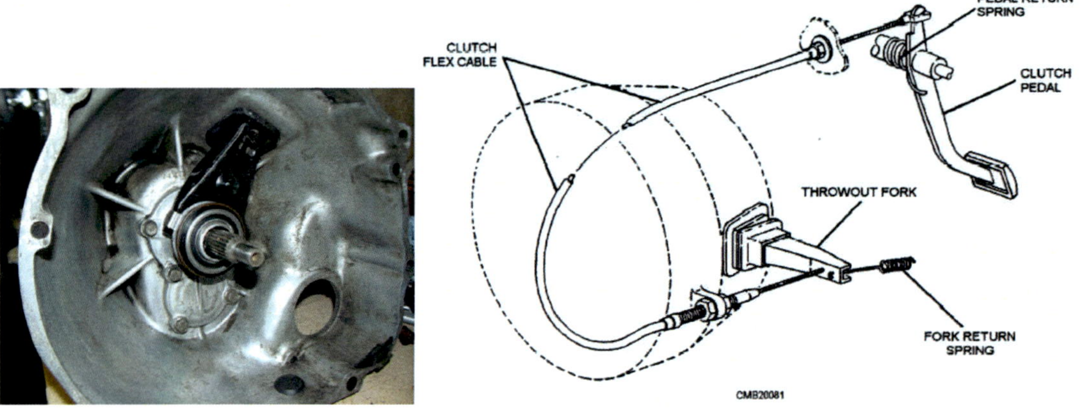

The Clutch Release Bearing

The clutch release bearing can be a thrust-type angular contact ball bearing. The bearing requires no periodic maintenance during its service life

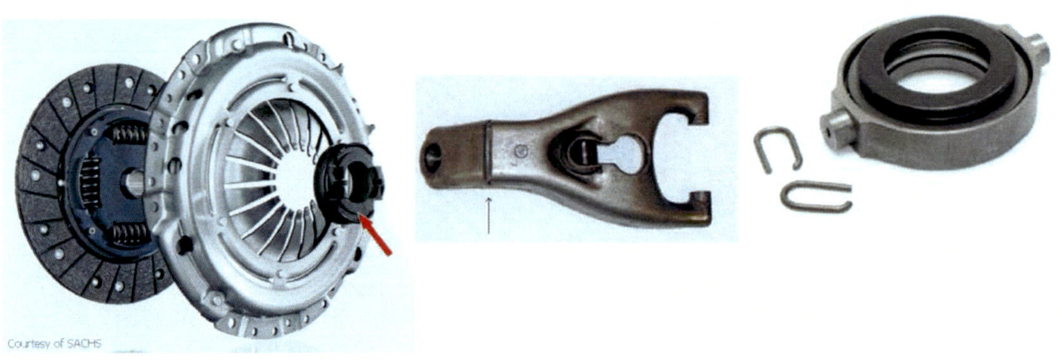

Clutch Release Bearing

- The main responsibility of a release bearing is to aid in the disengagement of the clutch. A clutch release bearing is attached to a transmission, where it pushes or sometimes pulls the springs inside of a pressure plate

Clutch Bearing

As the clutch pedal is pushed, the "fork" lever is moved down and the release bearing is pushed against the springs, relieving spring tension

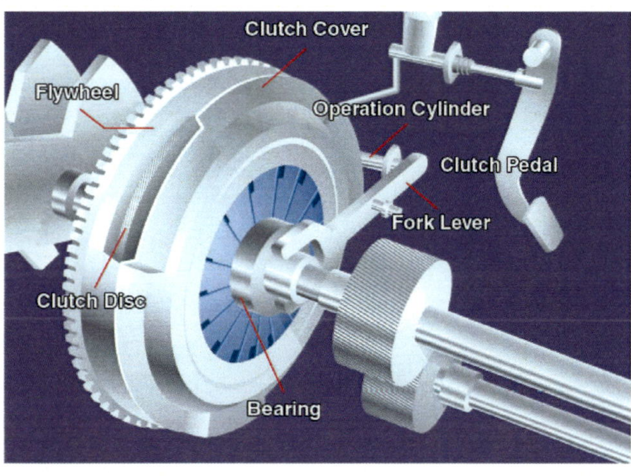

Hydraulic Clutch Bearing

The hydraulic bearing require less force applied at the pedal and otherwise produce around three times the pressure applied over the pressure plate in comparison with the mechanical system of fork and axial bearing

Mechanical Clutch Pedal

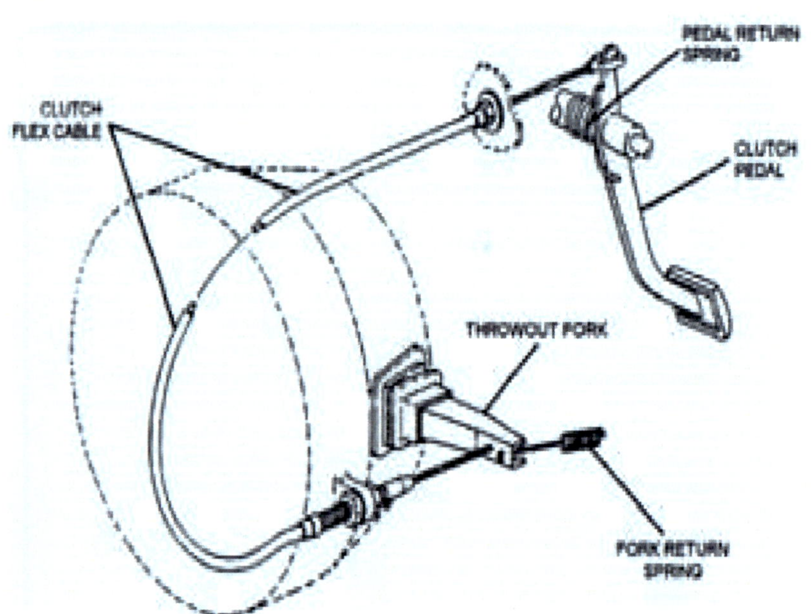

Hydraulic Clutch Pedal
(With mechanical Clutch Bearing Release)

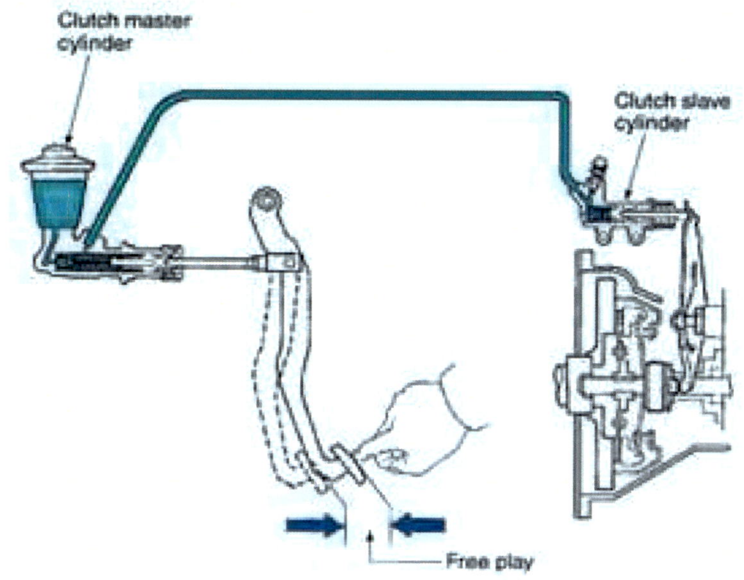

Hydraulic Clutch Bearing

(No Fork is required)

Manual Transmissions

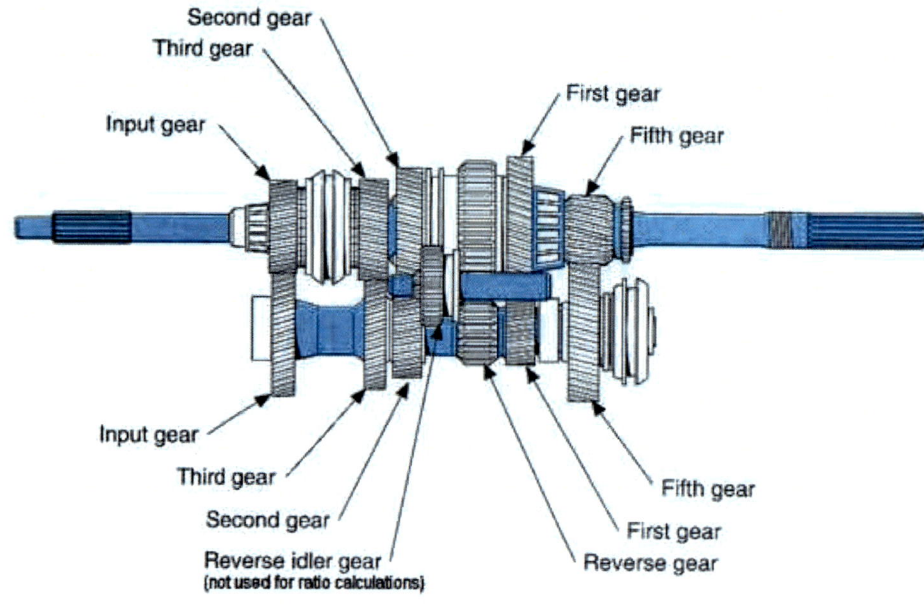

Manual transmission

Manual transmissions are characterized by gear ratios that are selectable by locking selected gear pairs to the output shaft inside the transmission

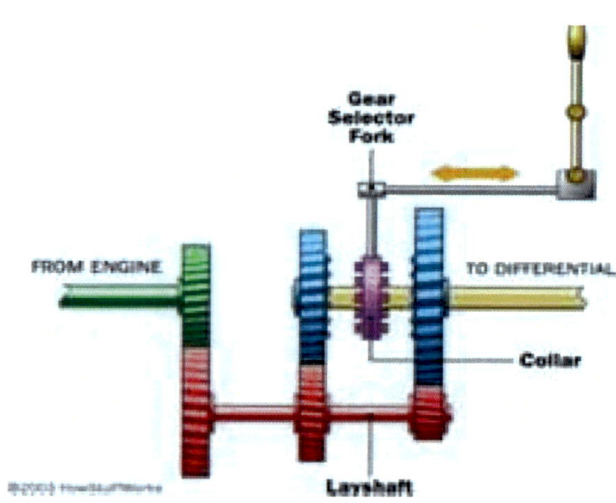

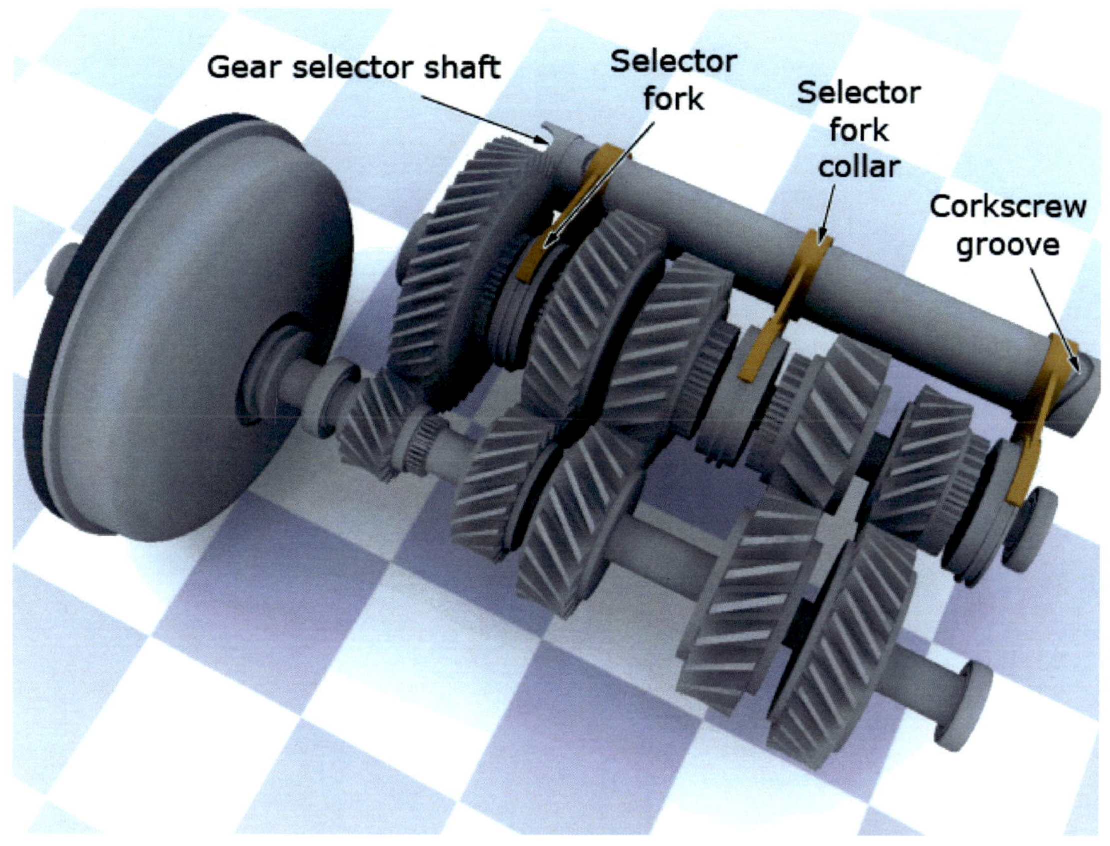

The Shafts

Typically, a manual transmission has three shafts: an **input shaft** (From Engine), a *countershaft* and an **output shaf**t. The countershaft is sometimes called a *layshaft*

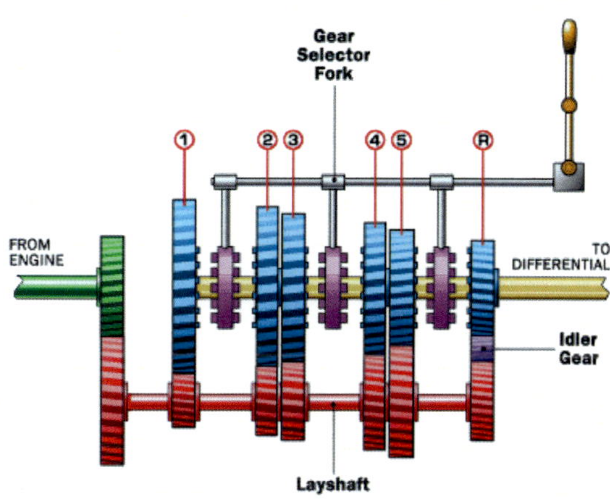

The Shafts

Pinions on the output shaft are not integral with the shaft. they only engage when the collar gear engages in one of their sides

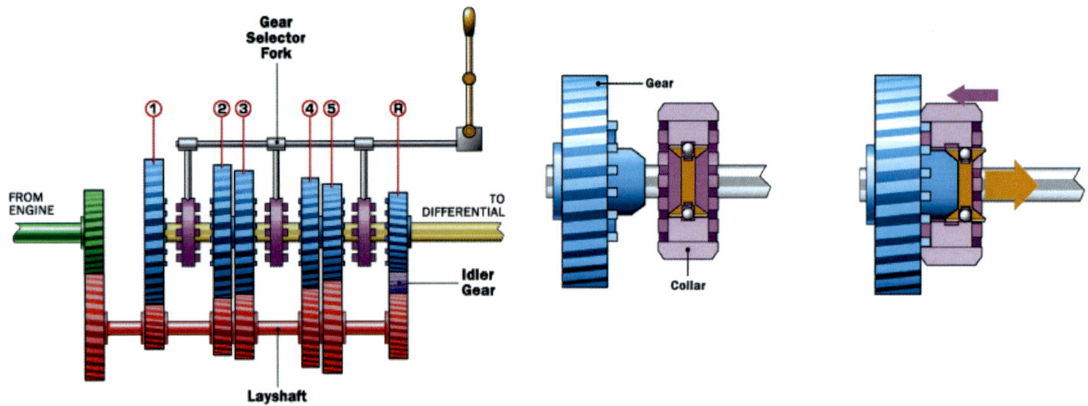

The Shafts

The transmission's input shaft **(Green)** has just one pinion gear, which drives the countershaft. Along the countershaft are mounted gears of various sizes, which rotate when the input shaft rotates.

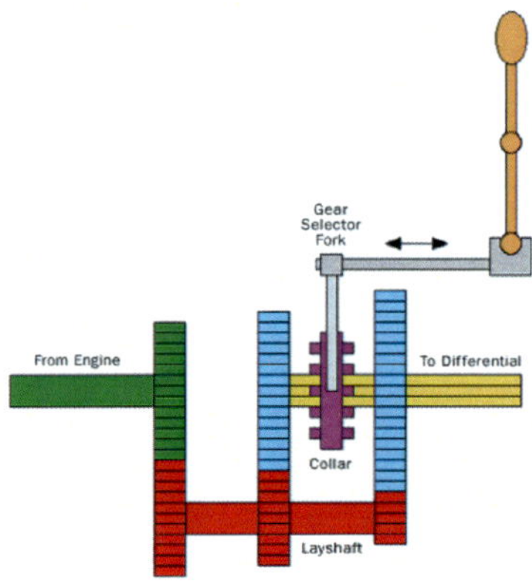

The Shafts

In some "marine gasoline transmission", the input and output shaft lie along the same line, and may in fact be combined into a single shaft within the transmission.

Automotive

Marine In-line

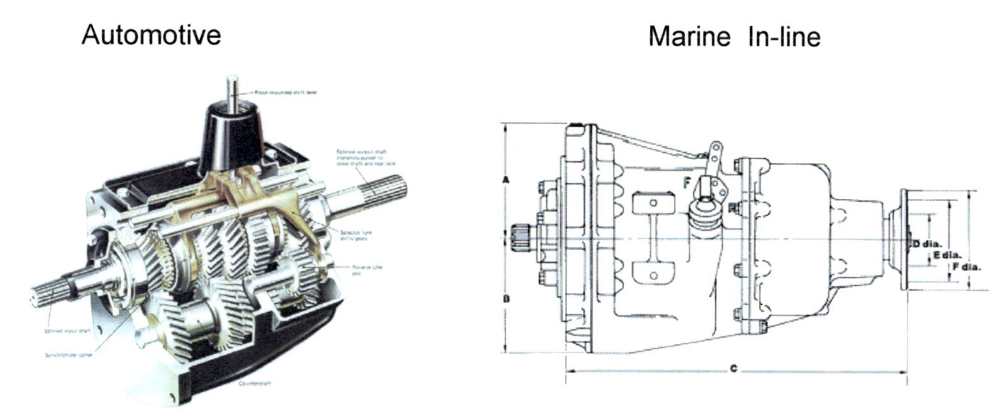

The Shafts

The output shaft **(Blue)** contain the gears correspond to the <u>forward speeds and reverse</u>. Each of the forward gears on the countershaft is permanently meshed with a corresponding gear on the output shaft.

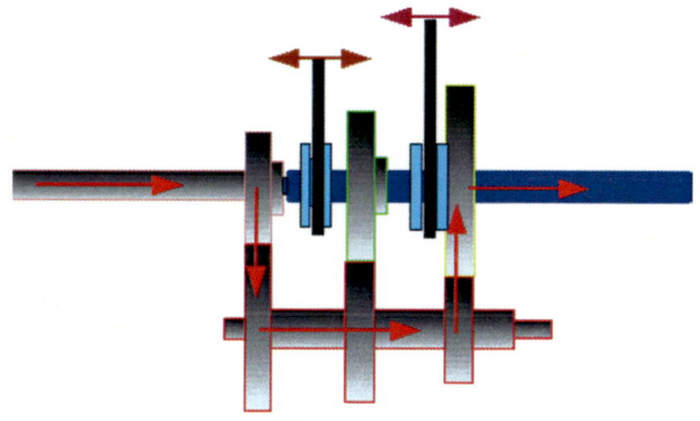

The Shafts

These driven gears are not rigidly attached to the output shaft: although the shaft runs through them, they spin independently of it, which is made possible by bearings in their hubs

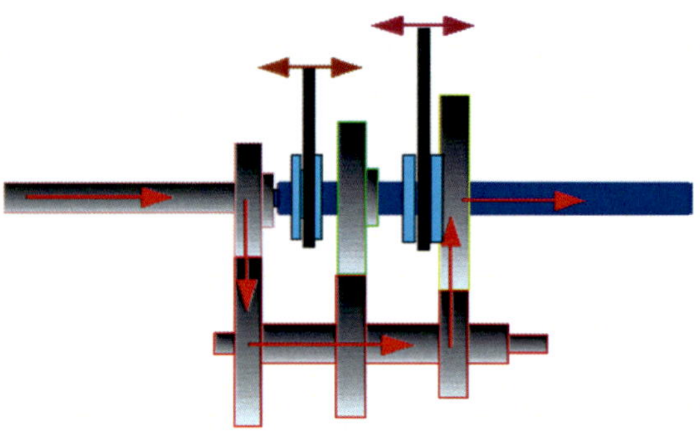

Collar Gear Operation

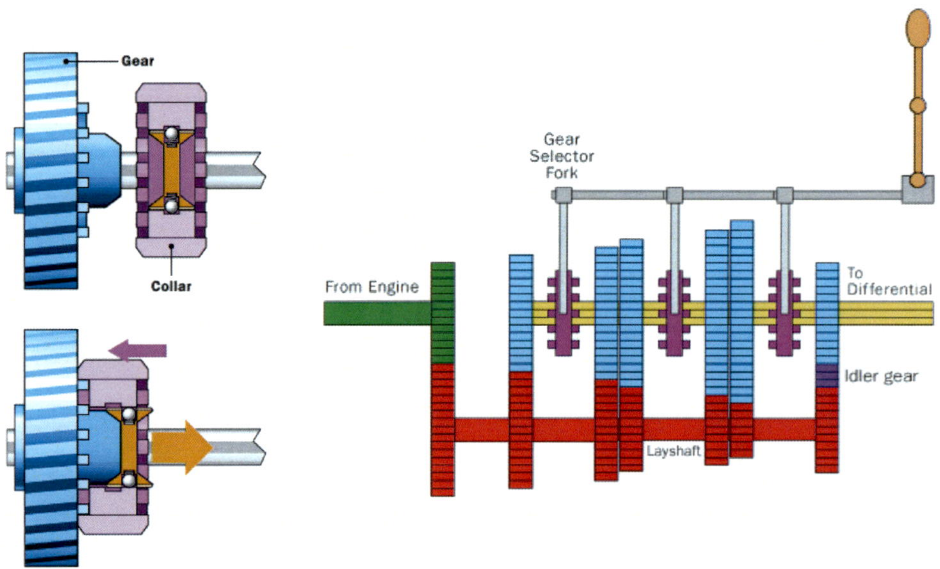

Manual transmission clutch
Neutral

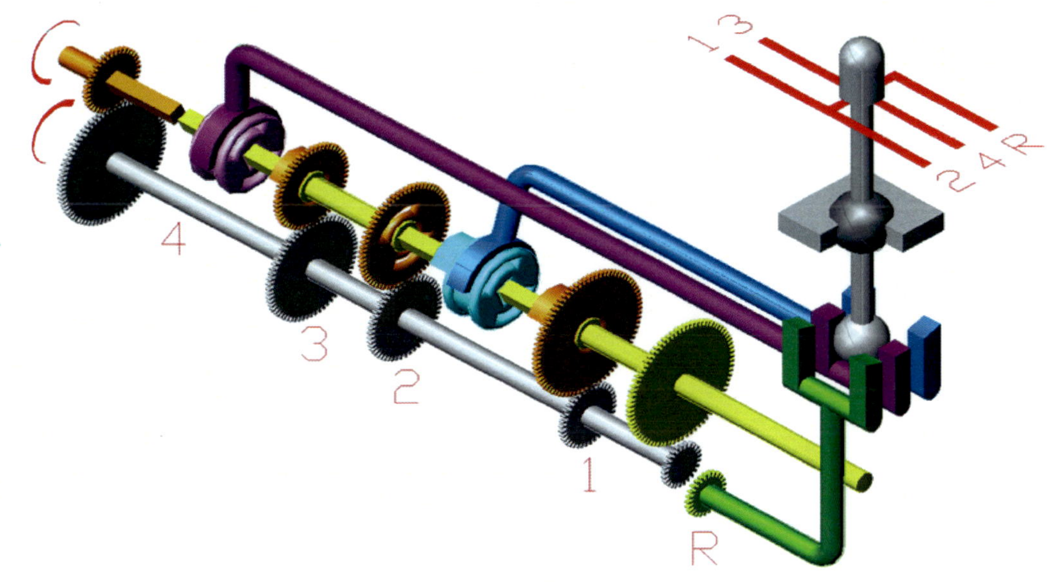

Output Shaft and Collar Gear

The output shaft has splines only where the **collar** moves . The gears of forward and reverse **(Purple)** turn freely on the shaft

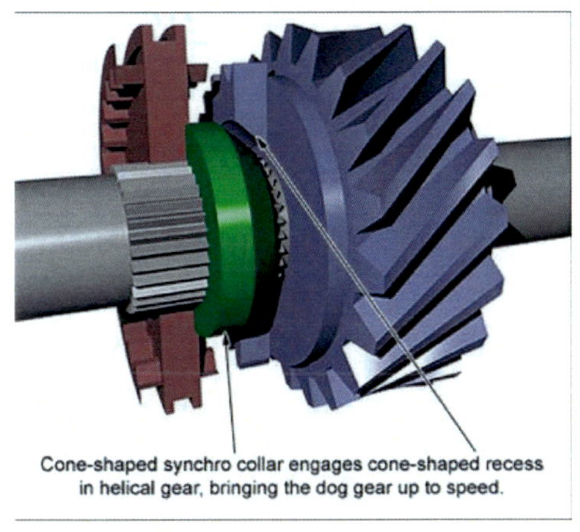
Cone-shaped synchro collar engages cone-shaped recess in helical gear, bringing the dog gear up to speed.

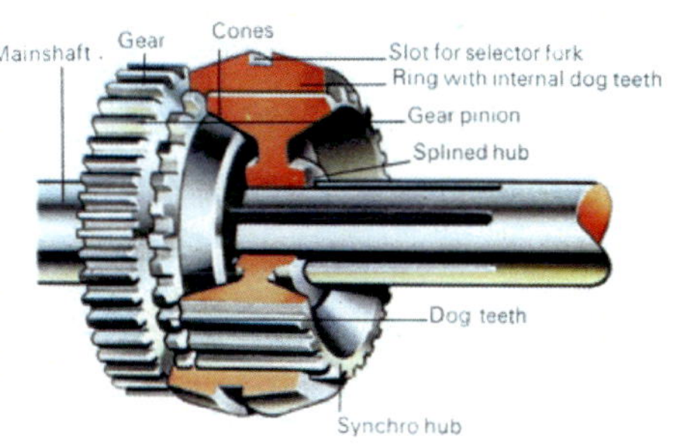

Dog clutch

- Locking the output shaft with a gear is achieved by means of a <u>dog clutch</u> selector.

- The dog clutch is a sliding selector mechanism which is splined to the output shaft, meaning that its hub has teeth that fit into slots (splines) on the shaft, forcing it to rotate with that shaft.

Dog clutch

The splines allow the selector to move back and forth on the shaft, which happens when it is pushed by a selector fork that is linked to the gear lever

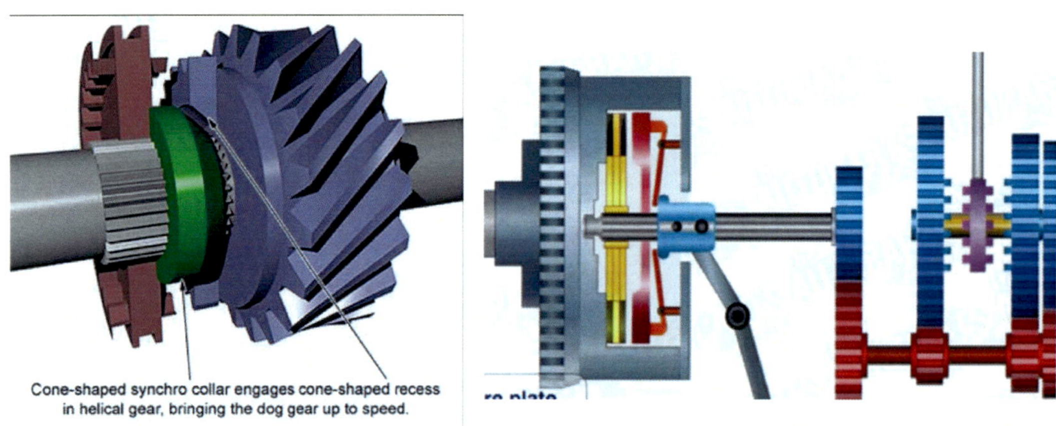

Dog clutch

The fork does not rotate, so it is attached to a collar bearing on the selector. The selector is typically symmetric: it slides between two gears and has a synchromesh and teeth on each side in order to lock either gear to the shaft

Dog clutch

The fork does not rotate, so it is attached to a collar bearing on the selector. The selector is typically symmetric: it slides between two gears and has a synchromesh and teeth on each side in order to lock either gear to the shaft

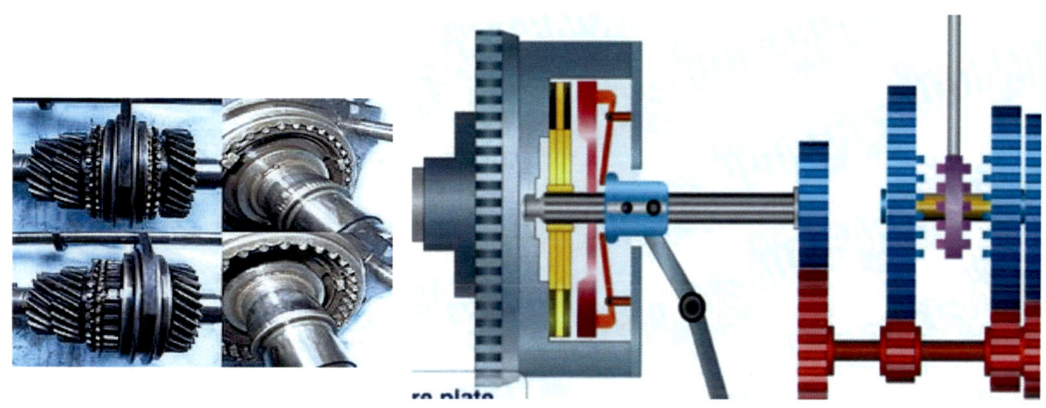

Dog clutch

The action of the gear selector is to lock one of the freely spinning gears (Blue) to the shaft that runs through its hub. Then shaft will spin together with that gear

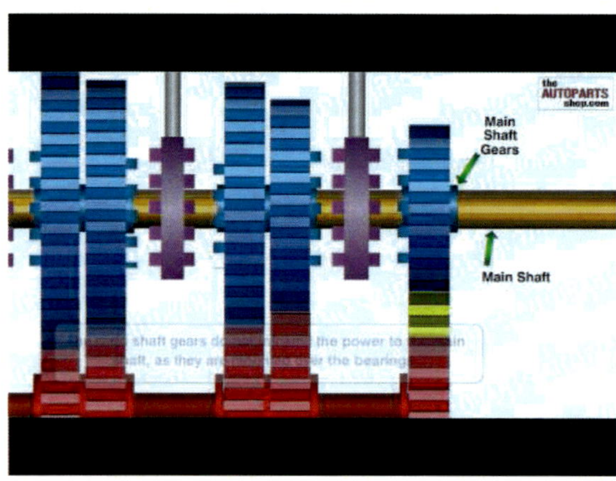

Dog clutch

The output shaft's speed relative to the countershaft is determined by the ratio of the two gears: the one permanently attached to the countershaft, and that gear's mate which is now locked to the output shaft.

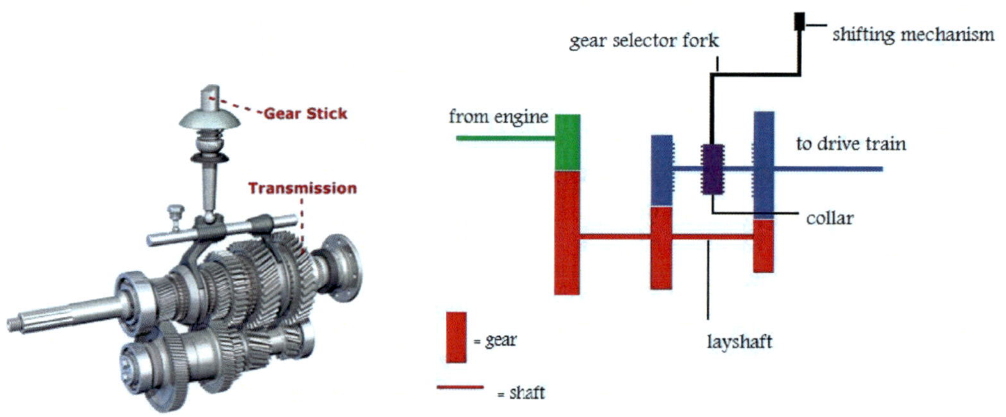

Synchronized transmission

Most modern engines are fitted with a synchronized gear box. Transmission gears are always in mesh and rotating, but gears on one shaft can freely rotate or be locked to the shaft.

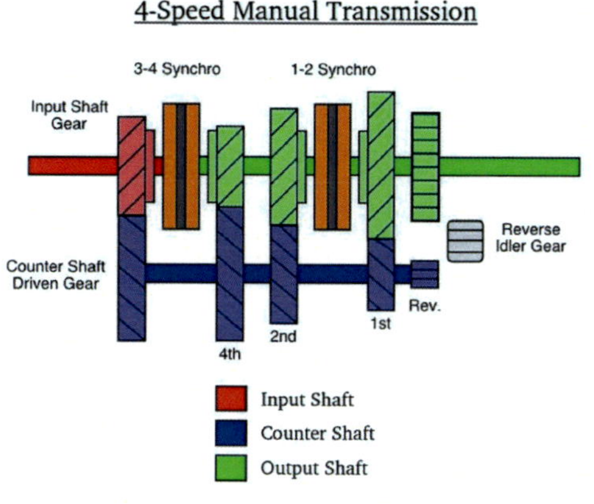

Synchronized transmission

The locking mechanism for a gear consists of a collar (or _dog collar_) on the shaft which is able to slide sideways so that teeth (or _dogs_) on its inner surface bridge two circular rings with teeth on their outer circumference : one attached to the gear, one to the shaft

How transmission works

Power enters the transmission through the **input shaft.** The input shaft is connected to the engine via the clutch, such that when the clutch is engaged, power goes straight from the engine to the input shaft of the transmission, and the crankshaft and input shaft rotate at the same speed

How transmission works

Just inside the housing of the transmission, the input shaft is connected to the countershaft (also known as the layshaft), by gears on both shafts, such that whenever the input shaft turns, so does the countershaft, and always at a fixed speed ratio

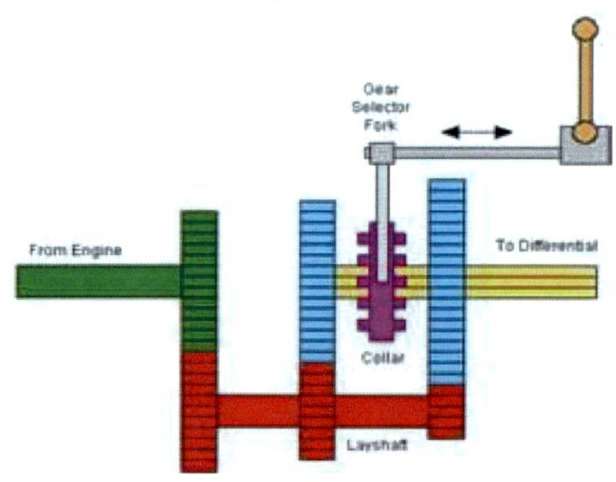

Forward First Gear

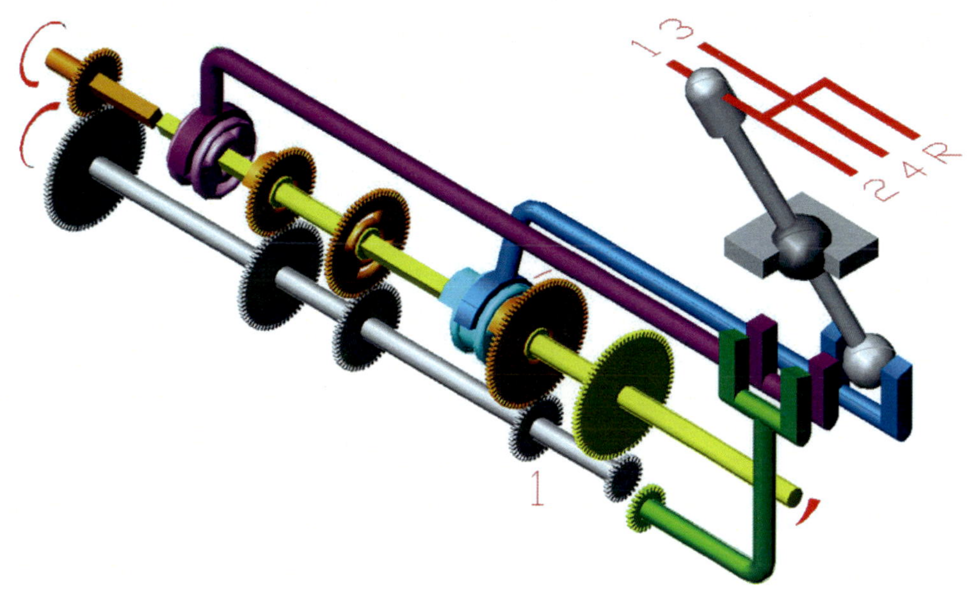

How transmission works

In addition to the gear that takes power from the input shaft, the countershaft has several gears on it, one for each of the car's "gears" including reverse

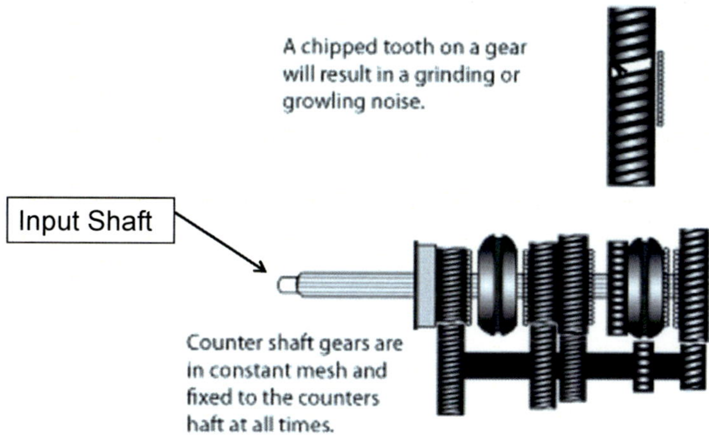

How transmission works

A third shaft called the <u>output shaft</u> runs parallel to the countershaft, and has freely-rotating <u>gears</u> that are mounted on bearings and rotate independently of the output shaft

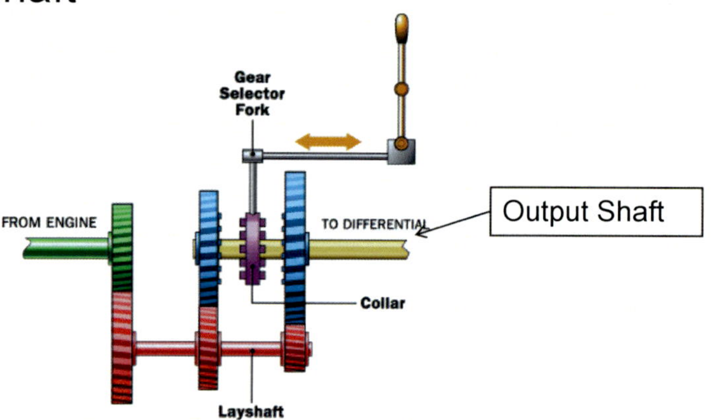

Forward Second Gear

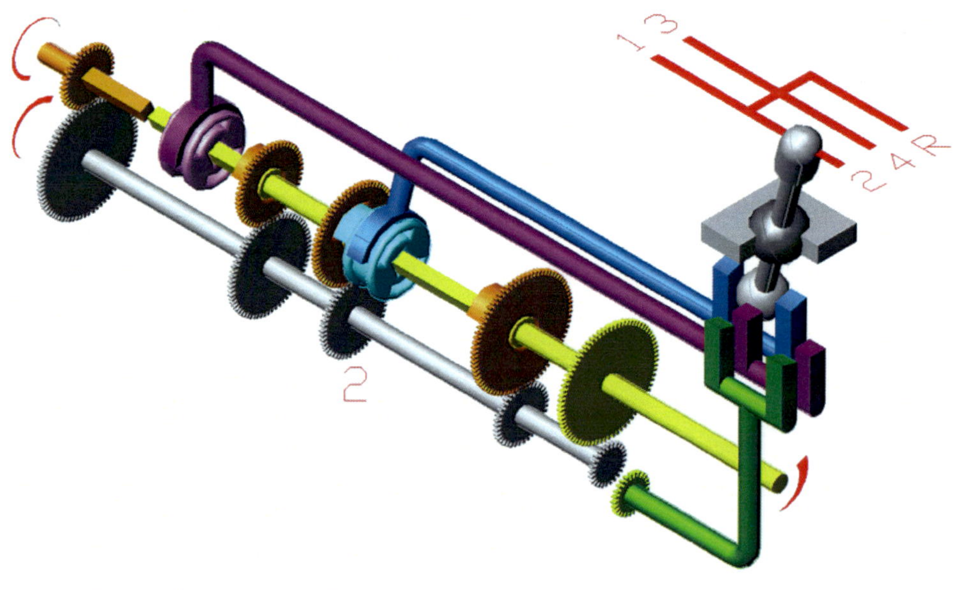

How transmission works

Each of these gears is paired with one of the countershaft gears, and is constantly in mesh with it. So actually the "freely-rotating" gears are constrained to rotate at a constant ratio with respect to the countershaft

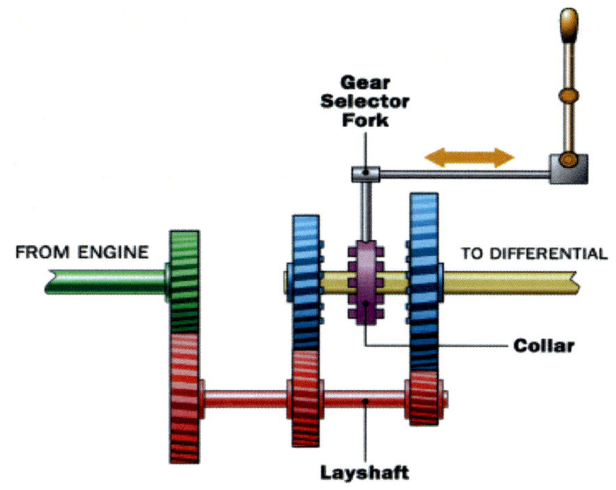

Reverse

Reverse is also a pair of gears: one gear on the countershaft and one on the output shaft. However, whereas all the forward gears are always meshed together, there is a gap between the reverse gears.

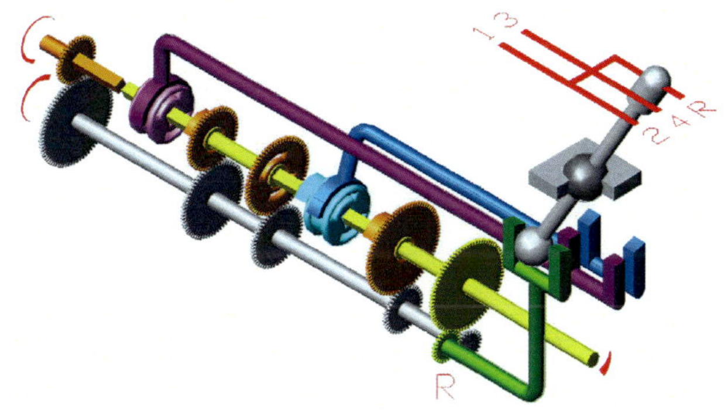

Reverse

What happens when reverse is selected is that a small gear, called an *idler gear* or *reverse idler*, is slid between them. The idler has teeth which mesh with both gears, and thus it couples these gears together and reverses the direction of rotation without changing the gear ratio.

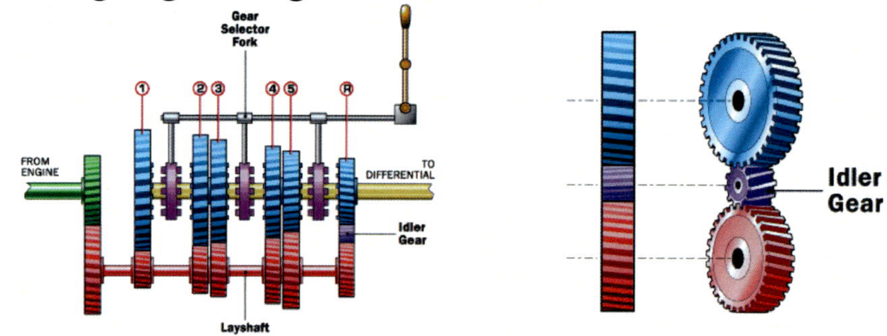

Reverse

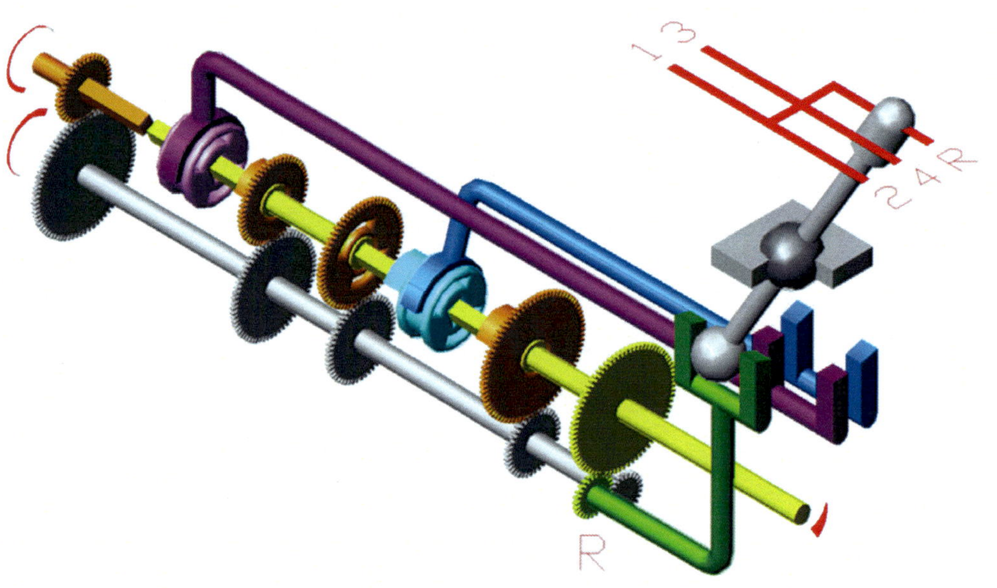

Reverse

- In other words, when reverse gear is selected, it is in fact *actual* gear teeth that are being meshed, with no aid from a synchronization mechanism.

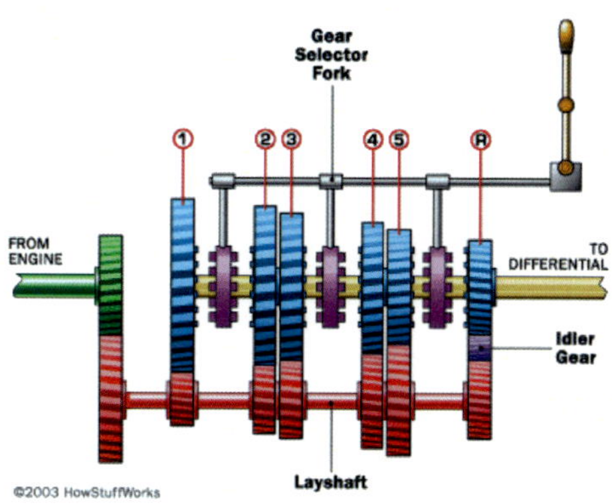

Reverse

- Whenever the clutch pedal is depressed to shift into reverse, the mainshaft continues to rotate because of its inertia.

- The resulting speed difference between mainshaft and reverse idler gear produces gear noise [grinding]. The reverse gear noise reduction system employs a cam plate which was added to the reverse shift holder.

Reverse

For this reason, the output shaft must not be rotating when reverse is selected: the engine must be in idle. In order that reverse can be selected without grinding even if the input shaft is spinning inertially, there may be a mechanism to stop the input shaft from spinning.

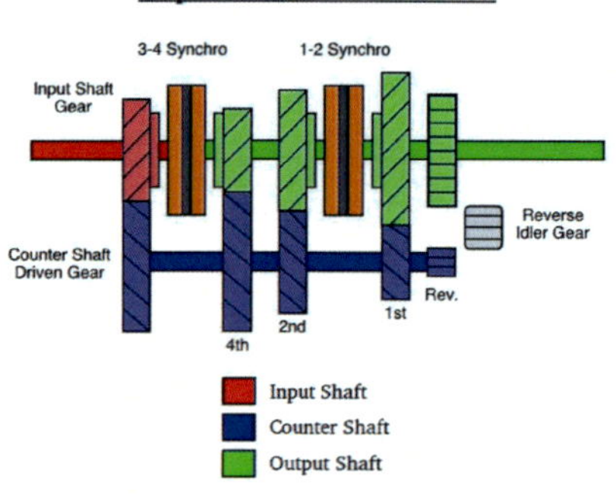

Synchronizer mechanism or *Synchromesh*

A modern dog clutch in a Diesel or Gas engine has a synchronizer mechanism or _synchromesh,_ which consists of a cone clutch and blocking ring.

Synchronized transmission

The gearshift lever manipulates the collars using a set of <u>linkages</u>, so arranged so that one collar may be permitted to lock only one gear at any one time; when "shifting gears," the locking collar from one gear is disengaged before that of another engaged

The Synchronizer Rings

The synchro ring rotates slightly due to the frictional torque from the cone clutch. In this position, the dog clutch is prevented from engaging.

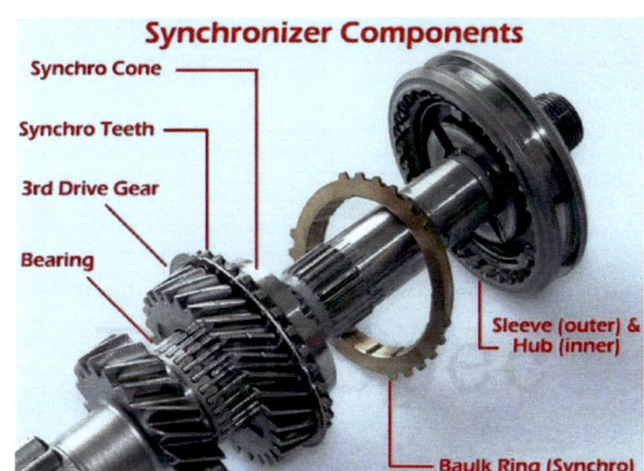

Synchronized transmission

The brass clutch ring gradually causes parts to spin at the same speed. When they do spin the same speed, there is no more torque from the cone clutch, and the dog clutch is allowed to fall in to engagement

The Synchronizer Rings

One collar often serves for two gears; sliding in one direction selects one transmission speed, in the other direction selects another.

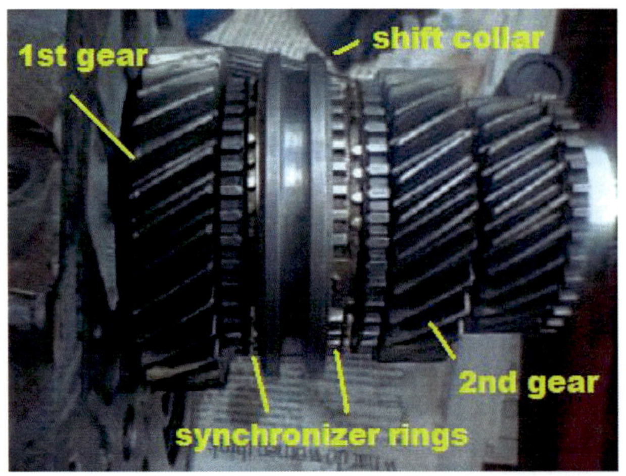

Synchronized transmission

In a synchromesh gearbox, to correctly match the speed of the gear to that of the shaft as the gear is engaged, the collar initially applies a force to a cone-shaped brass clutch attached to the gear, which brings the speeds to match prior to the collar locking into place

Synchromesh

If the teeth, the so-called dog teeth, make contact with the gear, but the two parts are spinning at different speeds, the teeth will fail to engage and a loud grinding sound will be heard as they clatter together

Synchromesh

Before the teeth can engage, the <u>cone clutch</u> engages first which brings the selector and gear to the same speed using friction. Moreover, until synchronization occurs, the teeth are prevented from making contact, because further motion of the selector is prevented by a *blocker* (or *baulk*) ring.

The Synchronizer Rings

The collar is prevented from bridging the locking rings when the speeds are mismatched by synchro rings

Chapter 4
Automatic Transmission

Automatic Transmission

Automatic transmissions contain mechanical systems, hydraulic systems, electrical systems and computer controls, all working together in perfect harmony which goes virtually unnoticed until there is a problem

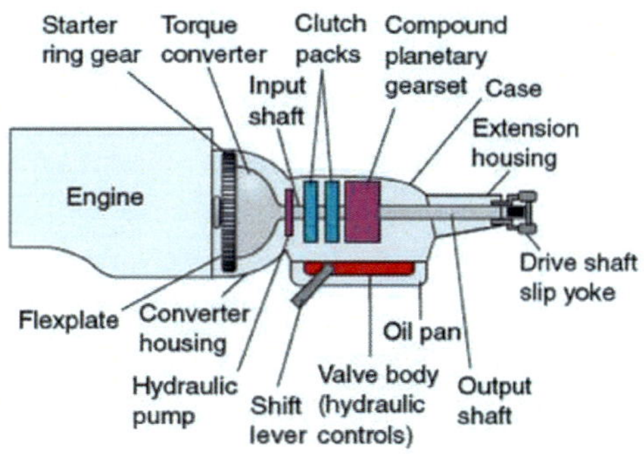

Transmission Components

- **Planetary Gear Sets** which are the mechanical systems that provide the various forward gear ratios as well as reverse

- The Hydraulic System which uses a special transmission fluid sent under pressure by an Oil Pump through the Valve Body to control the Clutches and the Bands in order to control the planetary gear sets.

- The **Torque Converter** which acts like a clutch to allow the vehicle to come to a stop in gear while the engine is still running

Transmission Components

Transmission Components

- The **Governor** and the **Modulator** or **Throttle Cable** that monitor speed and throttle position in order to determine when to shift.
  - On newer vehicles, shift points are controlled by a computer which directs electrical solenoids to shift oil flow to the appropriate component at the right instant
- **Seals and Gaskets** are used to keep the oil where it is supposed to be and prevent it from leaking out

Hydraulic Principle

The predominant form of automatic transmission is hydraulically operated; using a torque converter, and a set of planetary gears to provide a range of gear ratios

Torque Converter Vs. Mechanical Clutch

An automatic Transmission uses a torque converter instead of clutch to manage the connection between the transmission gearing and the engine

The Torque Converter

Torque Converter

The torque converter normally takes the place of a mechanical <u>clutch</u> (Inside of a Bell-housing) in a vehicle with an automatic transmission, allowing the load to be separated from the power source

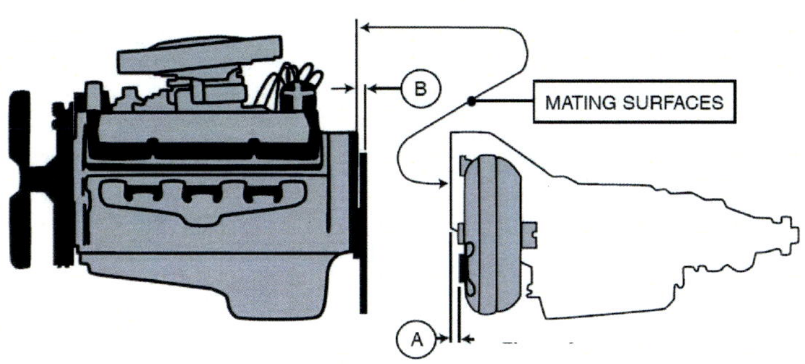

Torque Converter

In a torque converter there are at least three rotating elements: the impeller, which is mechanically driven by the <u>prime mover</u>; the turbine, which drives the <u>load</u>; and the stator, which is interposed between the impeller and turbine so that it can alter oil flow returning from the turbine to the impeller

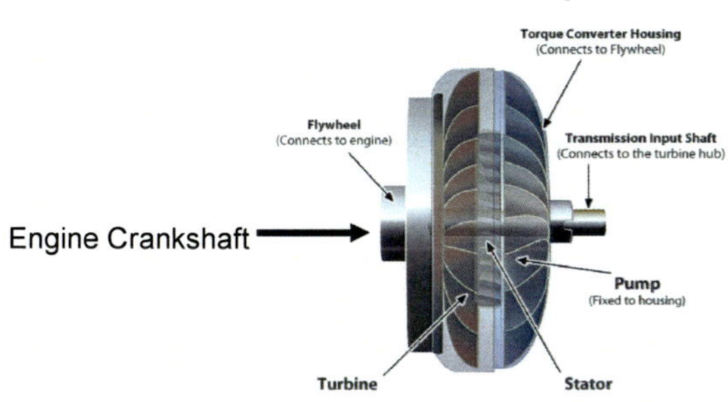

Pump

The pump is mounted directly to the converter housing which in turn is bolted directly to the engine's crankshaft and turns at engine speed

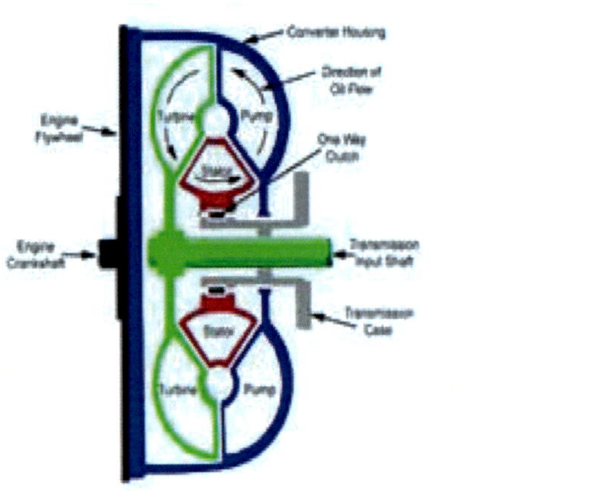

Turbine

- **The turbine** is inside the housing and is connected directly to the input shaft of the transmission providing power to move the output shaft

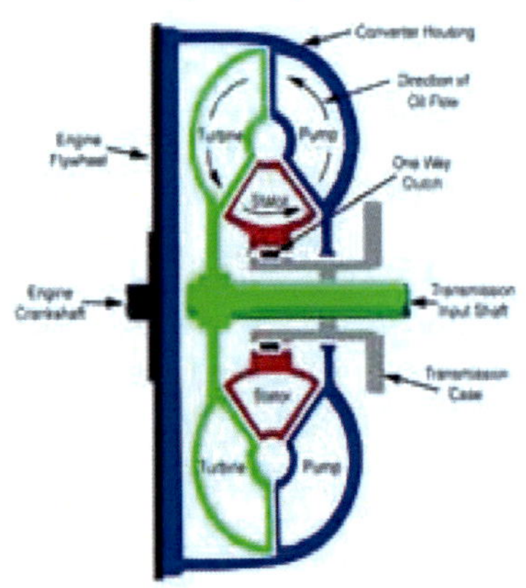

Stator

The Stator is mounted to a one-way clutch so that it can spin freely in one direction but not in the other

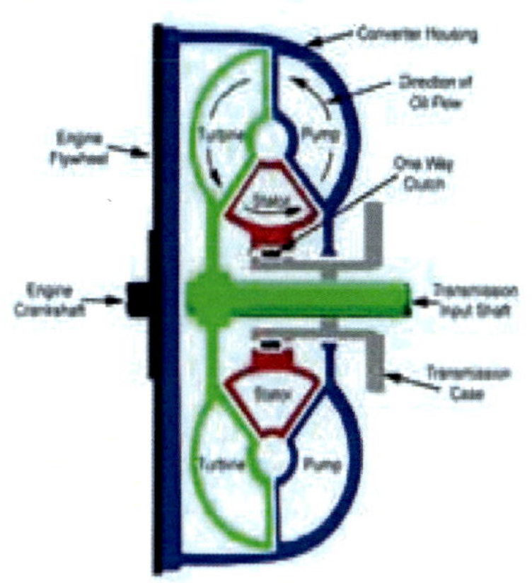

Operational Cycle

- With the engine running, transmission fluid is pulled into the pump section and is pushed outward by centrifugal force until it reaches the turbine section which starts it turning .

- The fluid continues in a circular motion back towards the center of the turbine where it enters the stator .

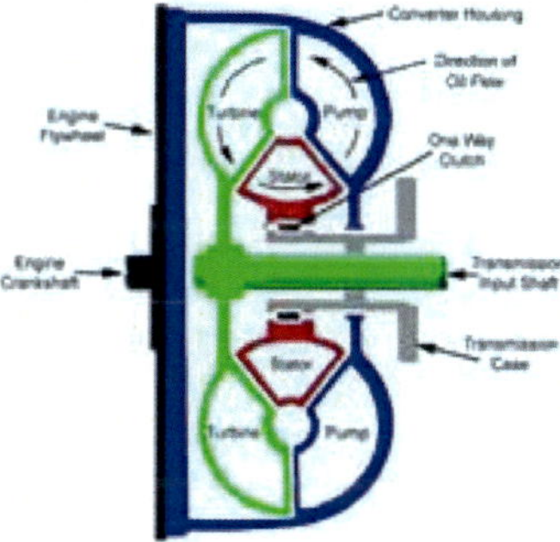

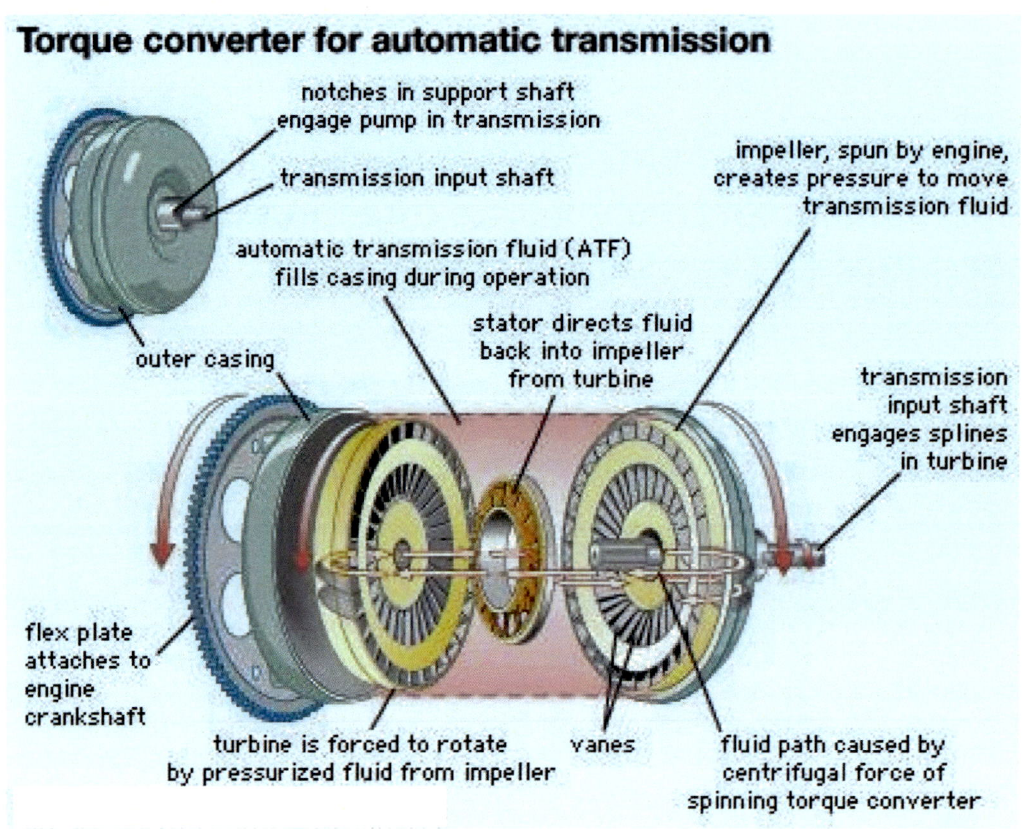

Operational Cycle

- If the turbine is moving considerably slower than the pump, the fluid will make contact with the front of the stator fins which push the stator into the one way clutch and prevent it from turning. With the stator stopped, the fluid is directed by the stator fins to re-enter the pump at a "helping" angle providing a torque increase

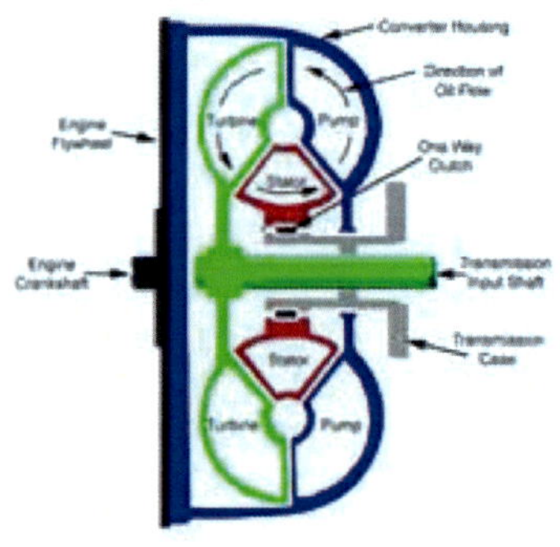

The Stator

The stator allows the smooth transition from standing still to motion of the vehicle and Also catches the fluid coming out of the turbine and redirects fluid flow

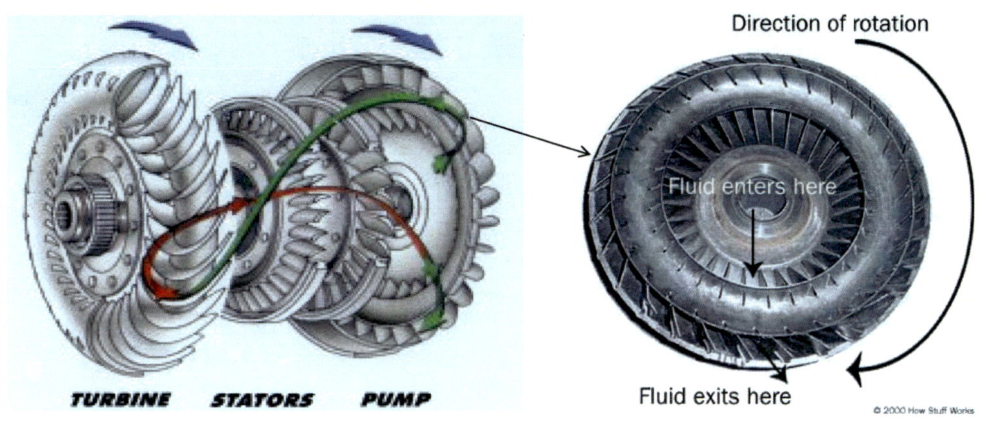

The Stator

The stator have two functions. Improve the flow of oil thru the turbine and help to return the fluid to the transmission

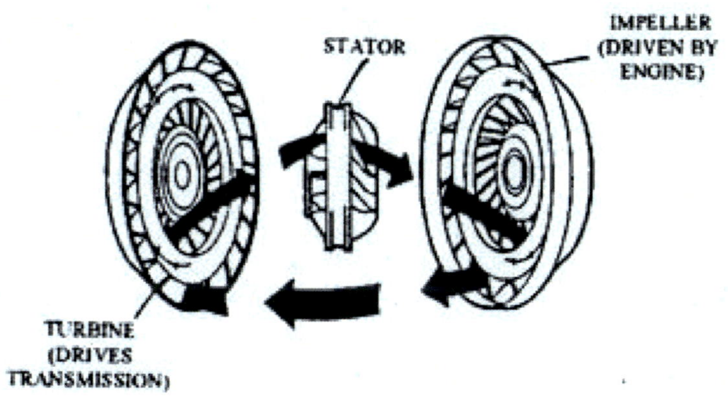

Stator / One Way Clutch

The stator is also called " One way clutch"

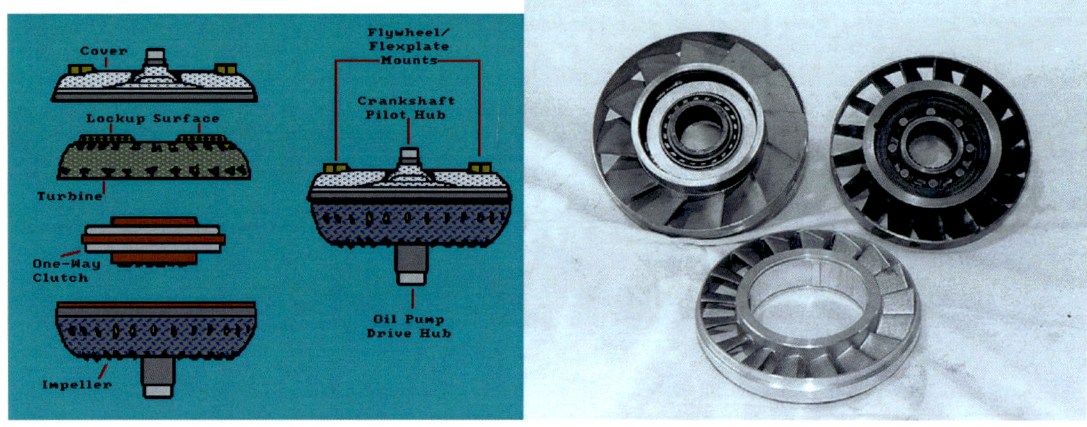

Operational Cycle

As the pump speed increase the fluid is send with high pressure into the turbine pushing it to make contact with the converter housing

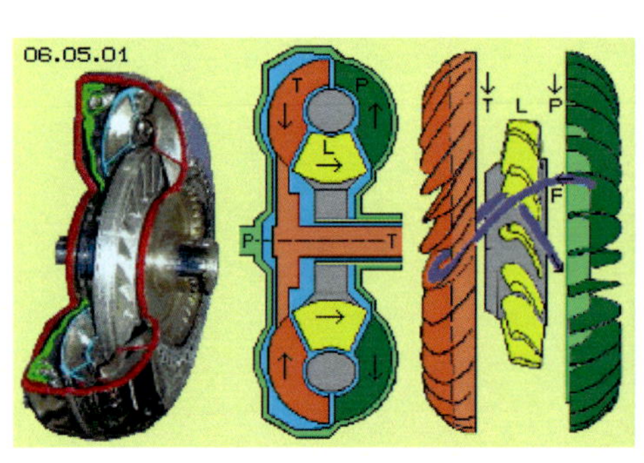

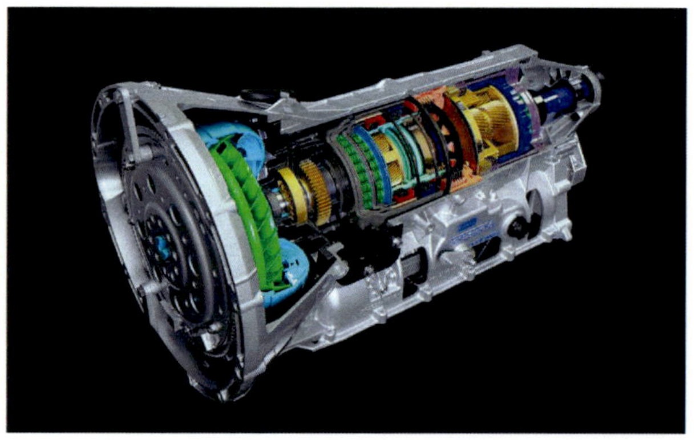

Operational Cycle

In this moment the turbine (Friction Plate) engage with the converter housing (Flywheel) and finally the Crankshaft is connected with the input shaft of the transmission

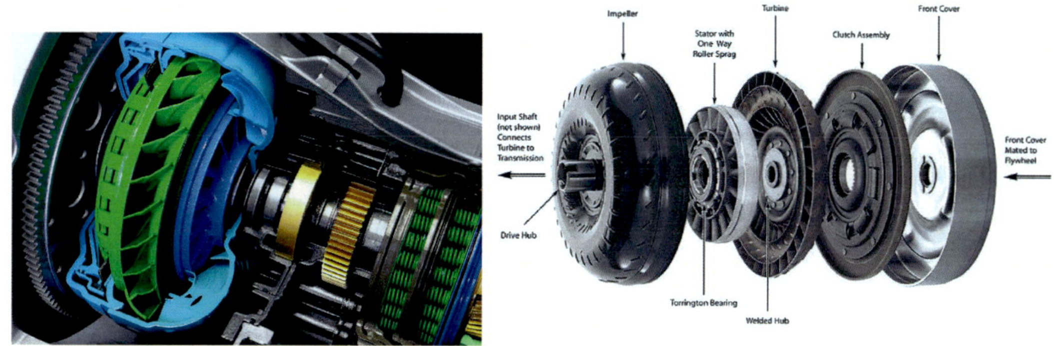

Shifting Problems?

If it is significantly more difficult to engage the transmission gear in forward as opposed to reverse, this indicates that the thrust washers are worn. The gear will require service or replacement

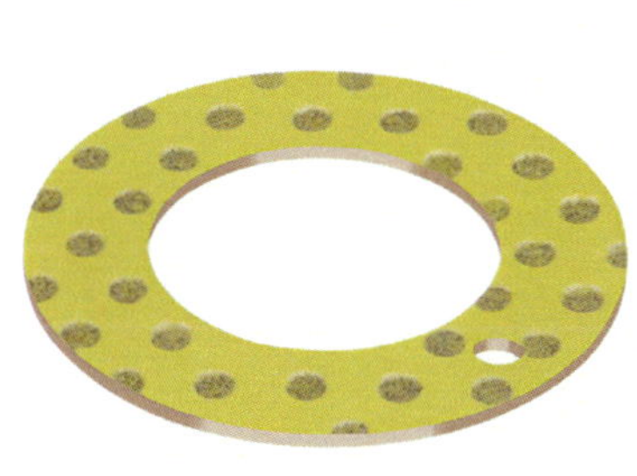

Operational phases

- A torque converter has three stages of operation
- **Stall**. The prime mover is applying power to the impeller but the turbine cannot rotate .
- **Acceleration**. The load is accelerating but there still is a relatively large difference between impeller and turbine speed .
- **Coupling**. The turbine has reached approximately 90 percent of the speed of the impeller

Stall Phase

- This stage of operation would occur when the driver has placed the transmission in gear but is preventing the vehicle from moving by continuing to apply the brakes.
- At stall, the torque converter can produce maximum torque multiplication if sufficient input power is applied (the resulting multiplication is called the *stall ratio*)

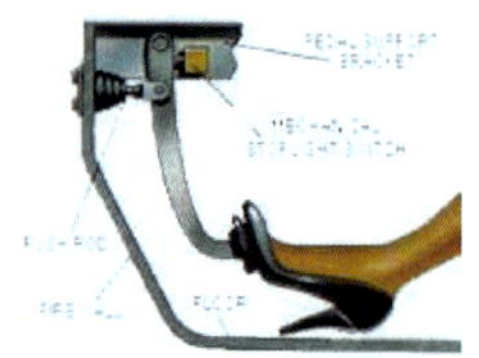

Acceleration Phase

- Under this condition, the converter will produce torque multiplication that is less than what could be achieved under stall conditions. The amount of multiplication will depend upon the actual difference between pump and turbine speed, as well as various other design factors

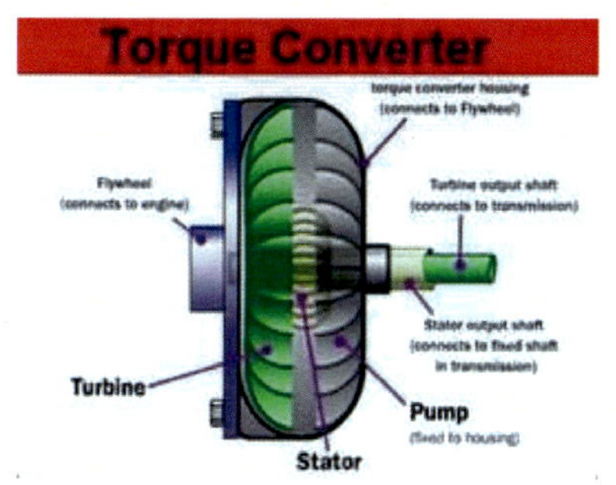

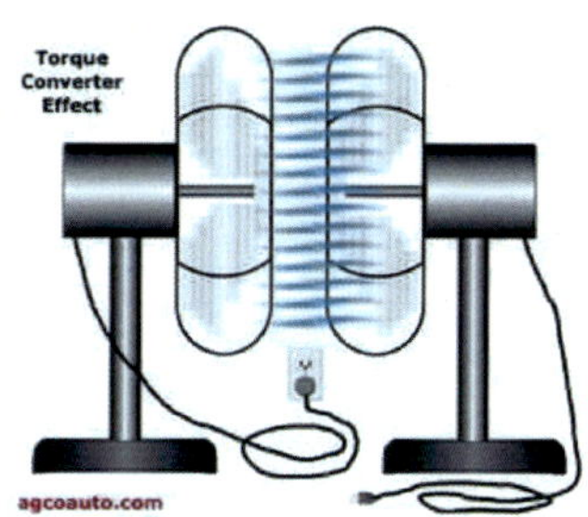

Acceleration Phase

- The principle behind a torque converter is like taking a fan that is plugged into the wall and blowing air into another fan which is unplugged. If you grab the blade on the unplugged fan, you are able to hold it from turning but as soon as you let go, it will begin to speed up until it comes close to the speed of the powered fan

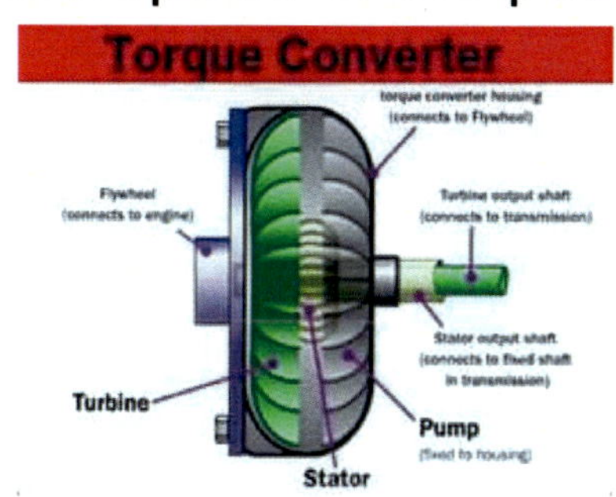

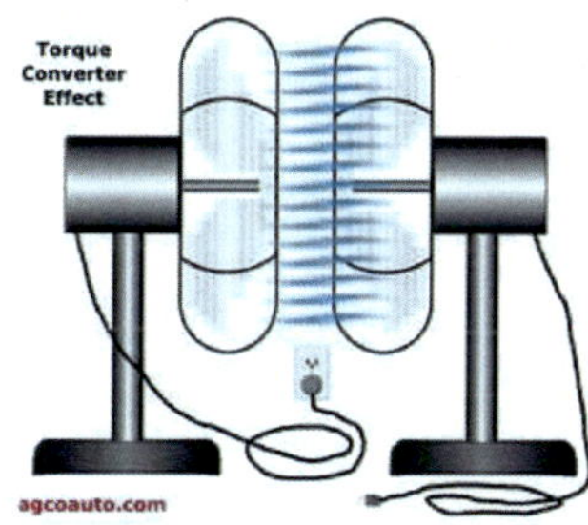

Coupling Phase

Torque multiplication has essentially ceased and the torque converter is behaving in a manner similar to a simple fluid coupling

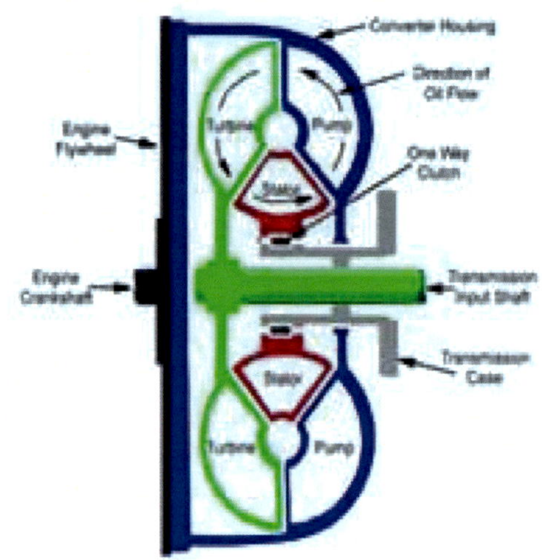

Efficiency

- A torque converter cannot achieve 100 percent coupling efficiency.

- The converter has zero efficiency at stall, generally increasing efficiency during the acceleration phase and low efficiency in the coupling phase .

- The loss of efficiency as the converter enters the coupling phase is a result of the turbulence and fluid flow interference generated by the stator, and as previously mentioned, is commonly overcome by mounting the stator on a one-way clutch .

Efficiency

- A torque converter cannot achieve 100 percent coupling efficiency.

- The converter has zero efficiency at stall, generally increasing efficiency during the acceleration phase and low efficiency in the coupling phase .

- The loss of efficiency as the converter enters the coupling phase is a result of the turbulence and fluid flow interference generated by the stator, and as previously mentioned, is commonly overcome by mounting the stator on a one-way clutch .

Torque Converter Problems

- ***Overheating***: It is the most common problem occurring in a torque converter. Persistent high levels of transmission slippage, hampers the converter's ability to dissipate heat

- ***Seizures in Stator Clutch***: Many times, a seizure in the stator clutch is caused by major loading and consequent disturbance of the clutch components

Torque Converter Problems

- ***Ballooning***: Persistent application under excessive loading, coupled with running the torque converter at a very high RPM can distort the shape of the converter's housing. This is called ballooning. It can cause the converter housing to burst or rupture .

- ***Deformation of Blade/ Turbine***: Excessive heating of the converter, can lead to the breakage or deformation of the turbine and pump. They can be separated from their hubs or annular rigs, or may break in fragments .

Torque Converter Reparation

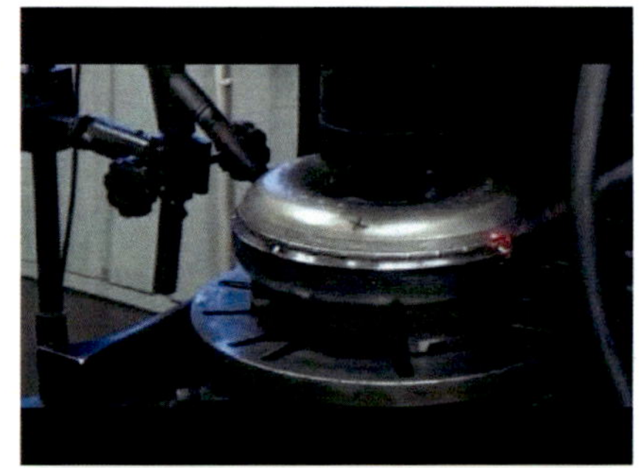

Planetary Gears

In a manual transmission, gears slide along shafts as you move the shift lever from one position to another, engaging various sized gears as required in order to provide the correct gear ratio

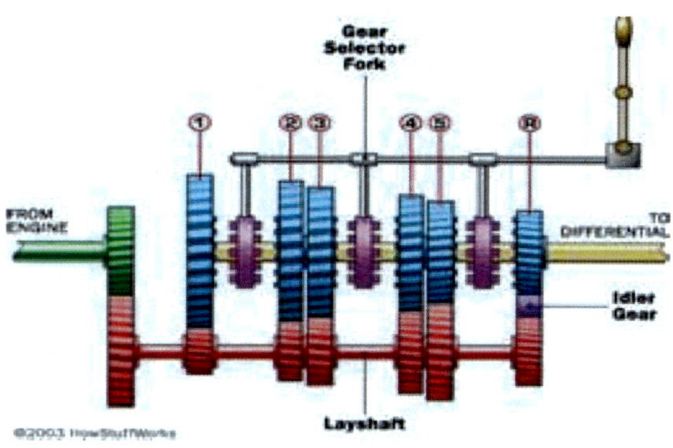

Planetary Gears

- In an automatic transmission, however, the gears are never physically moved and are always engaged to the same gears.

- This is accomplished through the use of planetary gear sets

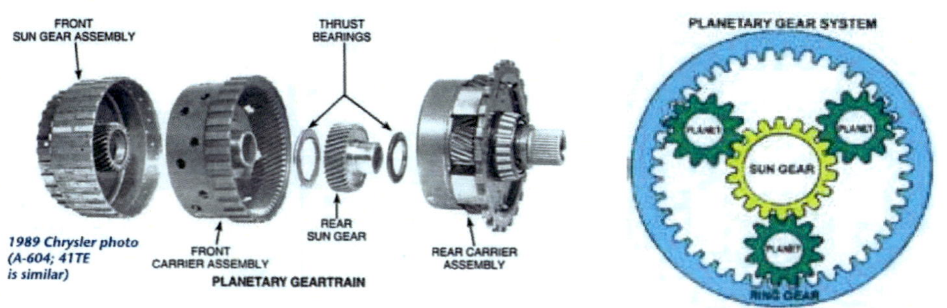

Planetary Gears

The basic planetary gear set consists of a <u>sun</u> gear, a ring gear and two or more planet gears, all remaining in constant mesh

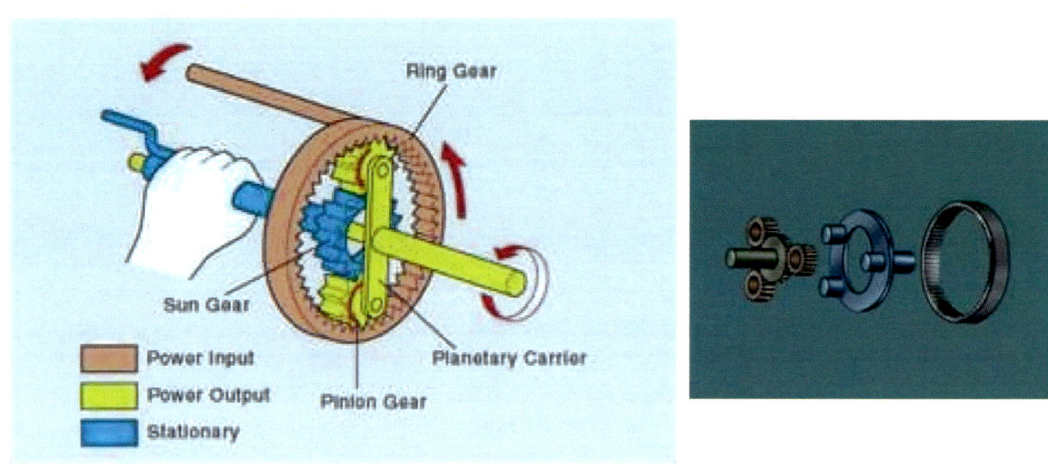

Planetary Gears

The planet gears are connected to each other through a common carrier which allows the gears to spin on shafts called "pinions" which are attached to the carrier

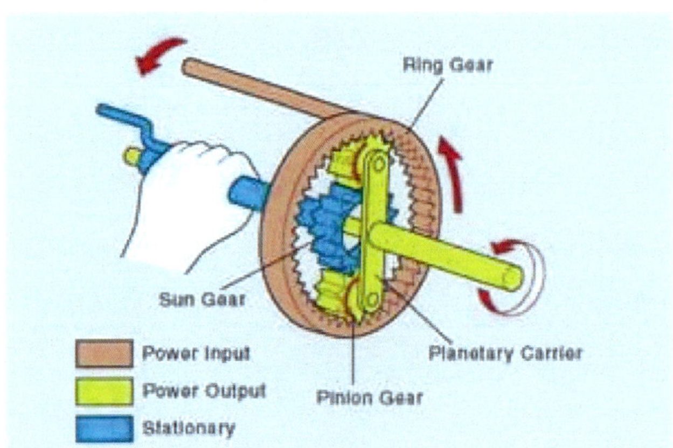

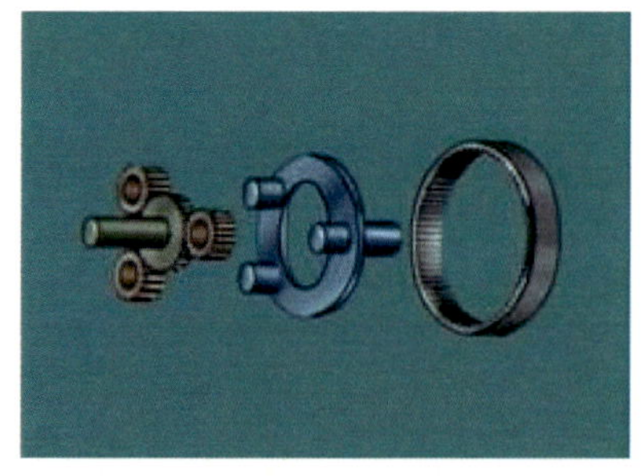

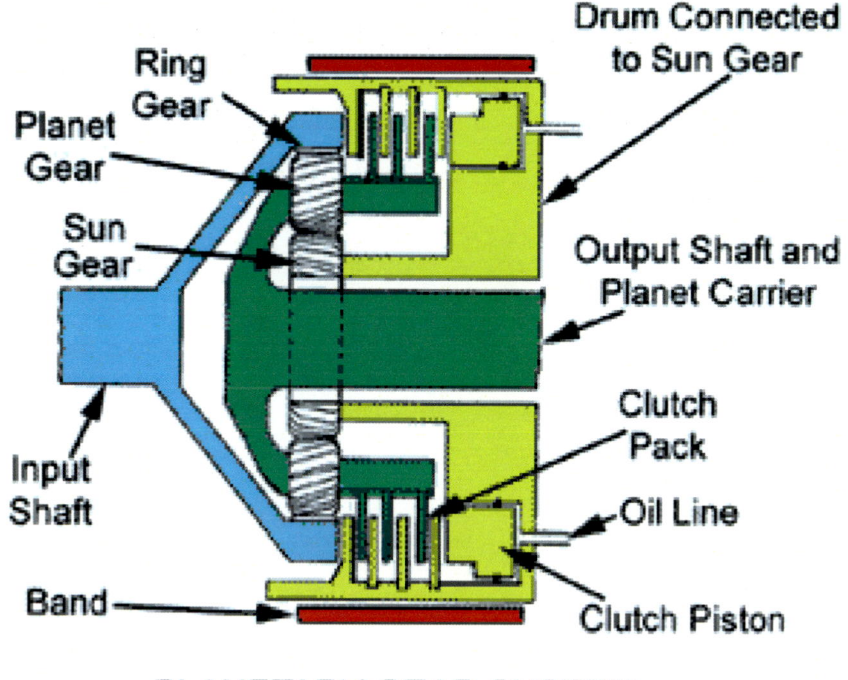

Planetary Gears

- One example of a way that this system can be used is by connecting the ring gear to the input shaft coming from the engine, connecting the planet carrier to the output shaft, and locking the sun gear so that it can't move

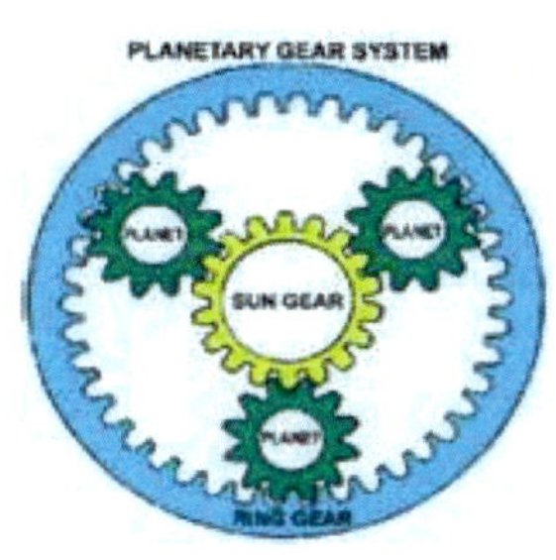

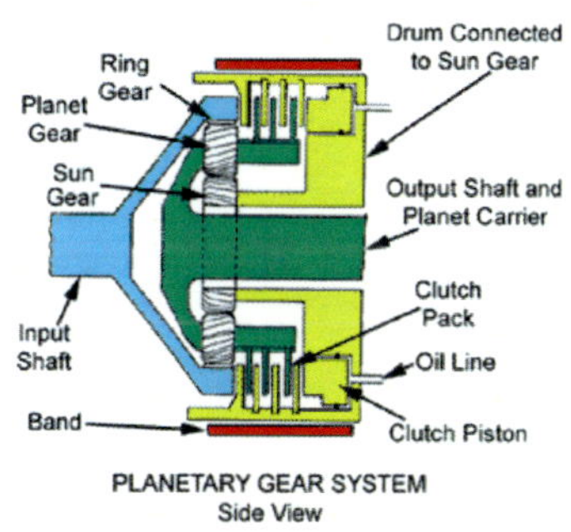

How The Planetary Gears works?

The input shaft is connected to the ring gear **(Blue)**, The Output shaft is connected to the planet carrier **(Green)** which is also connected to a "Multi-disk" clutch pack

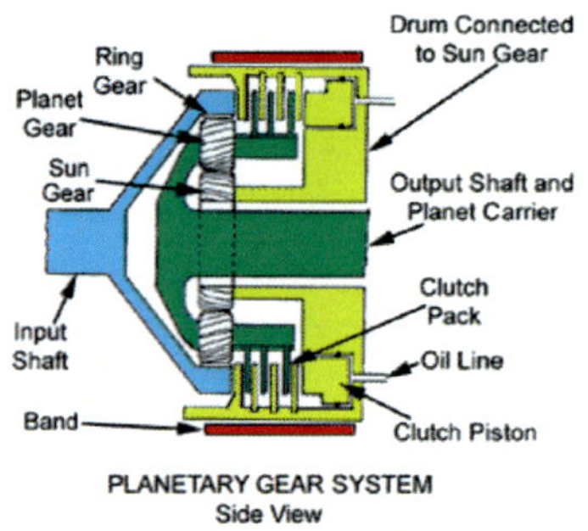

How The Planetary Gears Works?

The sun gear is connected to a drum **(yellow)** which is also connected to the other half of the clutch pack. Surrounding the outside of the drum is a band **(red)** that can be tightened around the drum when required to prevent the drum with the attached sun gear from turning.

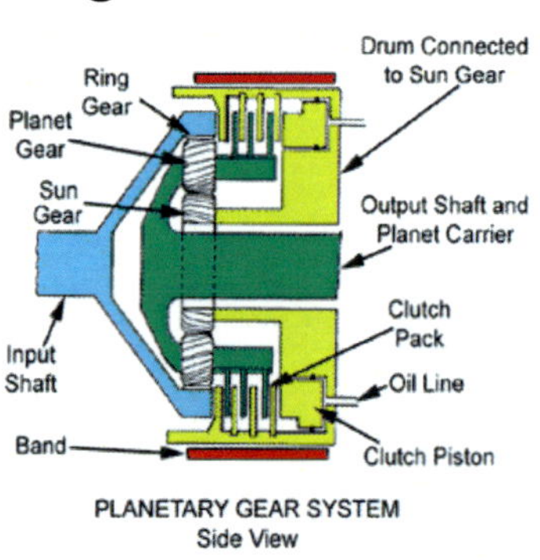

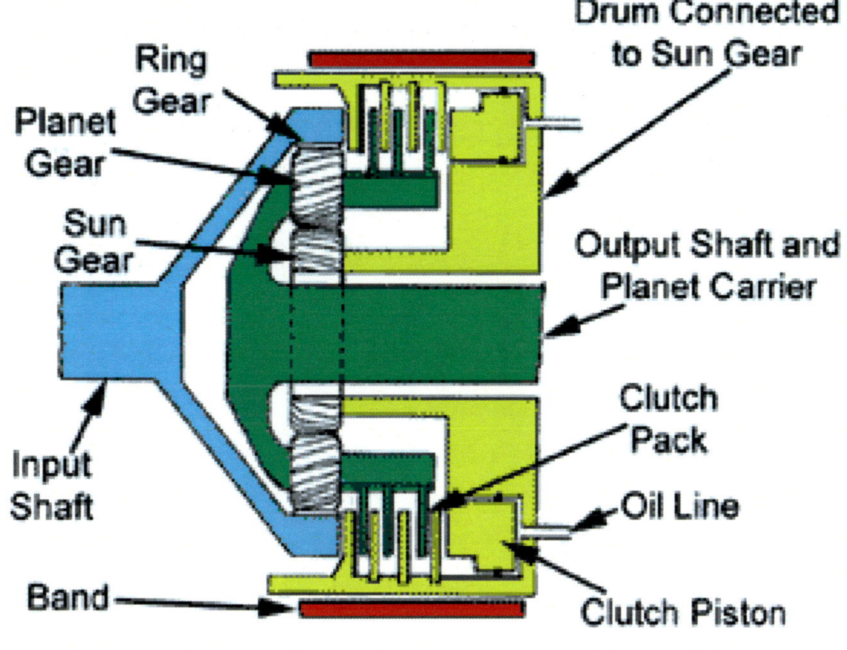

PLANETARY GEAR SYSTEM
Side View

First Gear

In this scenario, when we turn the **ring** gear, the planets will "walk" along the **sun** gear (which is held stationary) causing the **planet** carrier to turn the output shaft in the same direction as the input shaft but at a slower speed causing gear reduction (similar to a car in first gear)

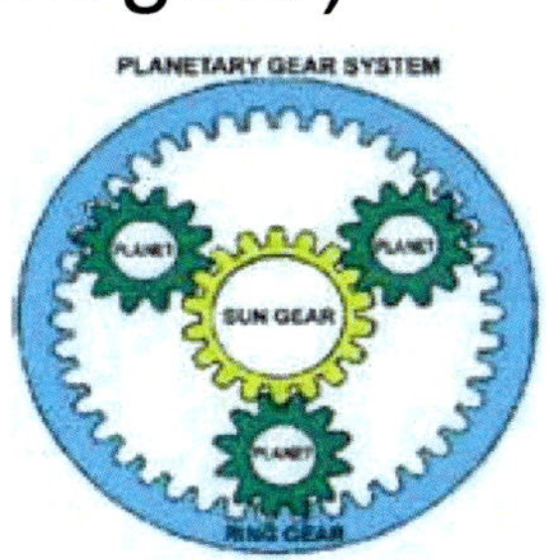

One Way Clutch

- A one-way clutch (also known as a "sprag" clutch) is a device that will allow a component such as ring gear to turn freely in one direction but not in the other

- A common place where a one-way clutch is used is in first gear when the shifter is in the drive position. When you begin to accelerate from a stop, the transmission starts out in first gear

High Gear

If we unlock the sun gear and lock any two elements together, this will cause all three elements to turn at the same speed so that the output shaft will turn at the same rate of speed as the input shaft. This is like a car that is in third or high gear

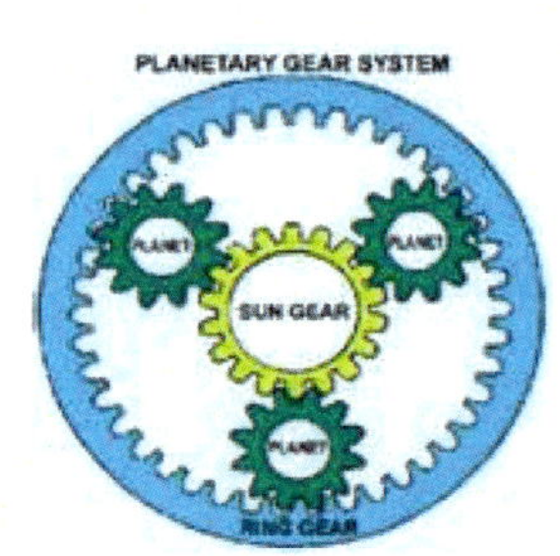

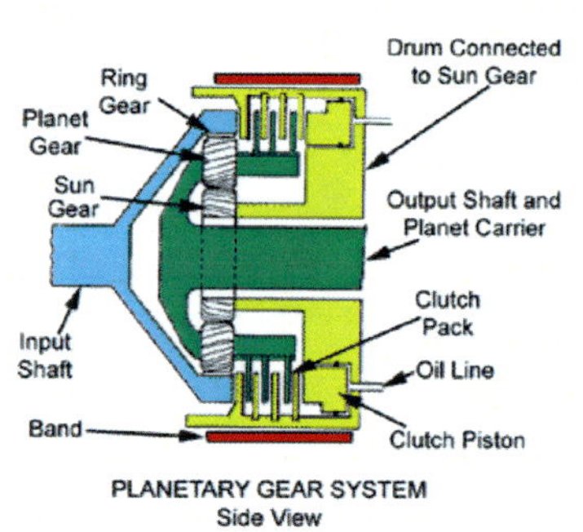

Second Gear

Some transmissions use the same sun gear with different combinations of solar and planet gears to obtain different outputs

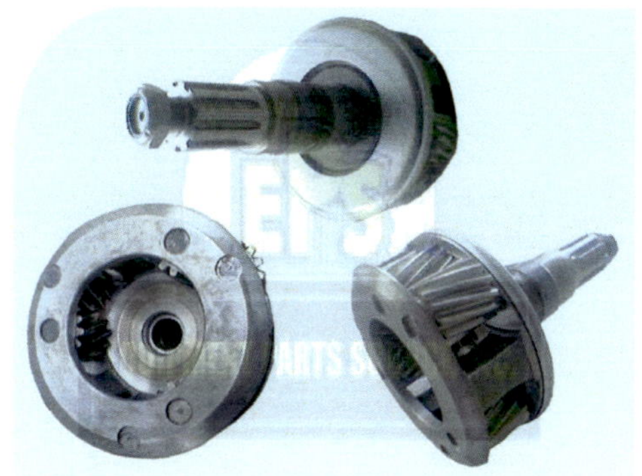

Reverse Gear

Another way that we can use a Planetary gear set is by locking the **planet** carrier from moving, then applying power to the **ring** gear which will cause the **sun** gear to turn in the opposite direction giving us **reverse gear**

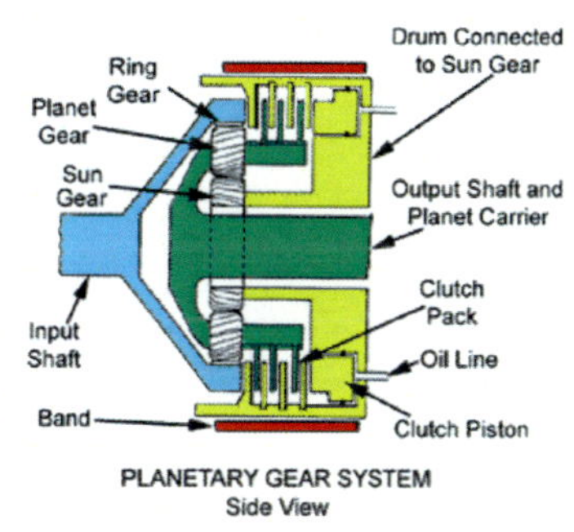

Sprag Clutch

is a device that will allow a component such as ring gear to turn freely in one direction but not in the other. This effect is just like that of a bicycle, where the pedals will turn the wheel when pedaling forward, but will spin free when pedaling backward

Sprag Clutch

A common place where a one-way clutch is used is in first gear when the shifter is in the drive position. When you begin to accelerate from a stop, the transmission starts out in first gear

Diagnosing One Way Clutch Problems

A common sign is a clicking sound from the transmission. If left alone, this will progress to complete loss of forward speeds

The engine revs very high, but the vehicle slows down when the driver tries to accelerate. After coming to a complete stop, the vehicle will once again start off

Servo Bands

A band is a steel strap with friction material bonded to the inside surface. One end of the band is anchored against the transmission case while the other end is connected to a servo

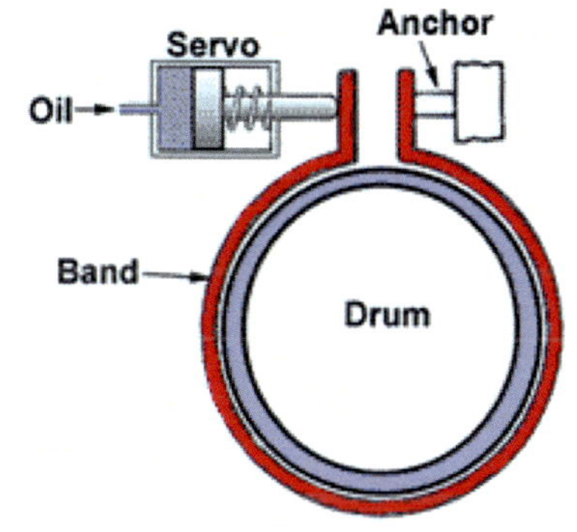

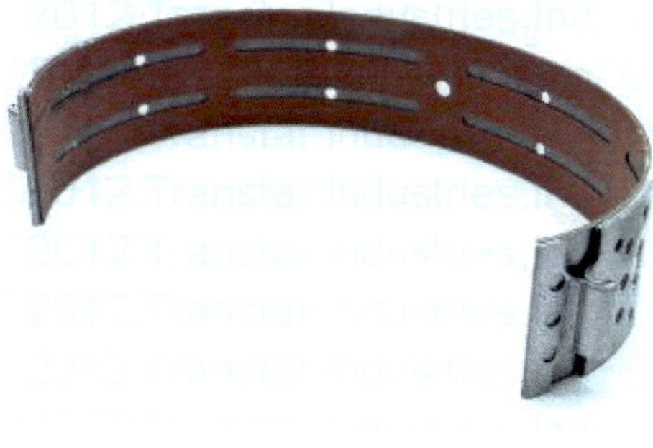

Bands

A band is a steel strap with friction material bonded to the inside surface. One end of the band is anchored against the transmission case while the other end is connected to a servo

Servo Band Brake

- One end of the band is anchored against the transmission case (See Arrow) while the other end is connected to a servo

- At the appropriate time hydraulic oil is sent to the servo under pressure to tighten the band around the drum to stop the drum from turning

Servo Bands

At the appropriate time hydraulic oil is sent to the servo under pressure to tighten the band around the drum to stop the drum from turning

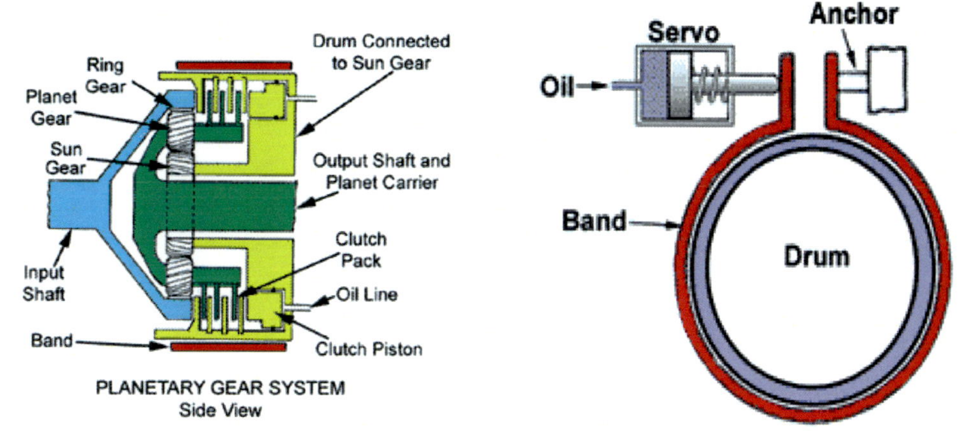

Multiplate Clutch Operation

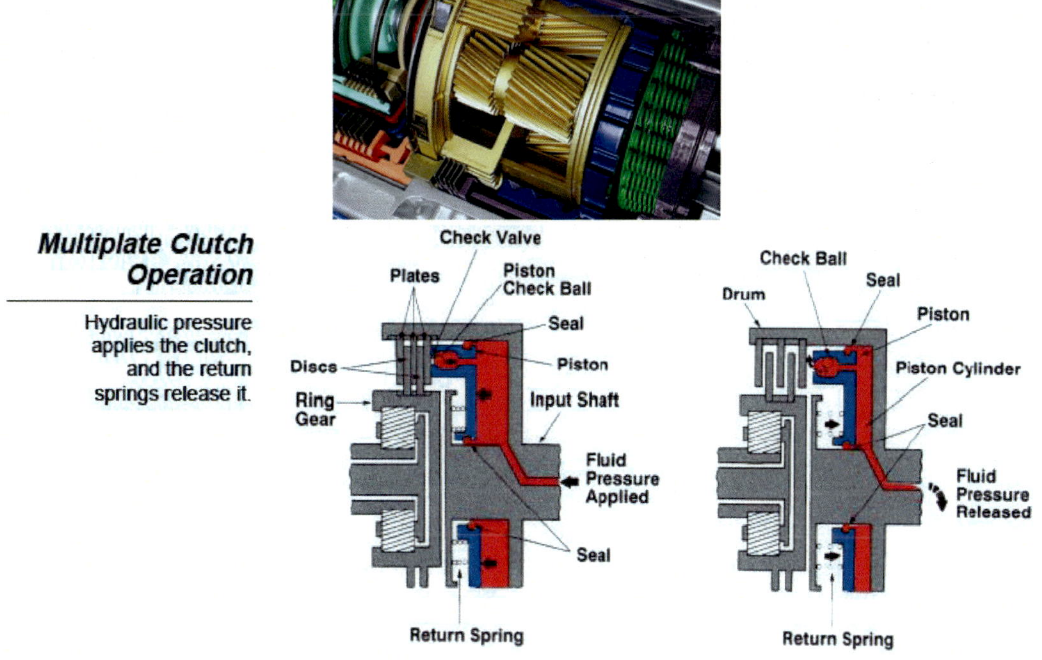

The Clutch Pack

A clutch pack consists of alternating disks (Half Steel and Half Friction material) that fit into groves on the inside of the drum and on the outer surface of the adjoining hub

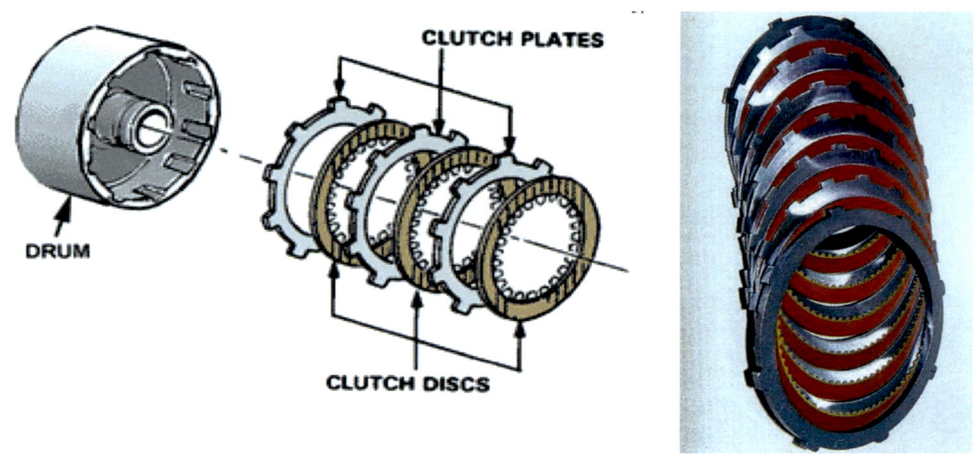

The Clutch Pack

The clutch pack is used, in this instance, to lock the planet carrier with the sun gear forcing both to turn at the same speed

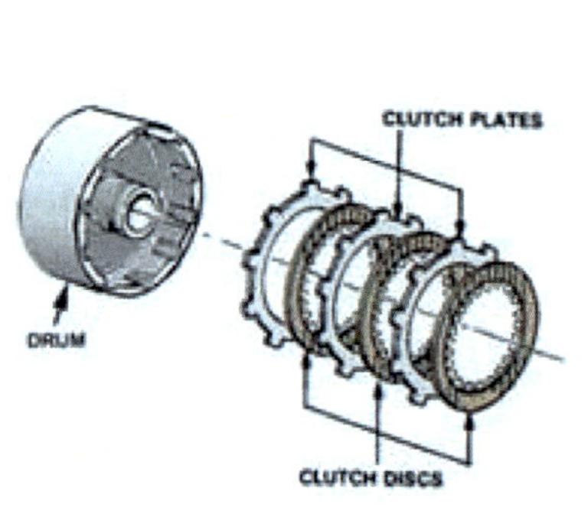

The Clutch Pack

The Clutch Pack

- Two sets of plates make up the clutch pack. These are driving plates and driven plates

- The inner driven plates have a friction material bonded to them and their splines mate with splines on the outside of a central hub, which is itself splined to the primary sun gear

The Clutch Pack

The Clutch Pack

- The friction material can be of treated paper or fibre and may have a grooved surface to assist in wiping oil from between the plates when the clutch engages

- The plates are placed alternately to make up the pack and are loaded into the clutch drum against the relatively thick pressure plate

Clutch Pack Parts

(Left) The transmissions' rear band. (Center) The forward, direct, and intermediate clutch packs. (Right) A look at the hard anodized, billet aluminum

The Clutch Pack

- The number of plates installed determines the torque capacity of the clutch. Installing more plates increases the torque capacity

- A centrifugal relief valve in the clutch piston releases fluid trapped in the cylinder when the clutch is released. This prevents partial application of the clutch which may be caused by centrifugal force acting on the fluid at high speed

Clutch Pack Assembly

Mock up of the clutch removal/installer to hold the intermediate brake drum, forward clutch and input shell assemblies together

Oil Pump

Automatic transmissions use what is called a gear pump. They usually contain two gears. The inner gear is driven by the torque converter hub (the hub is slotted to engage the inner gear)

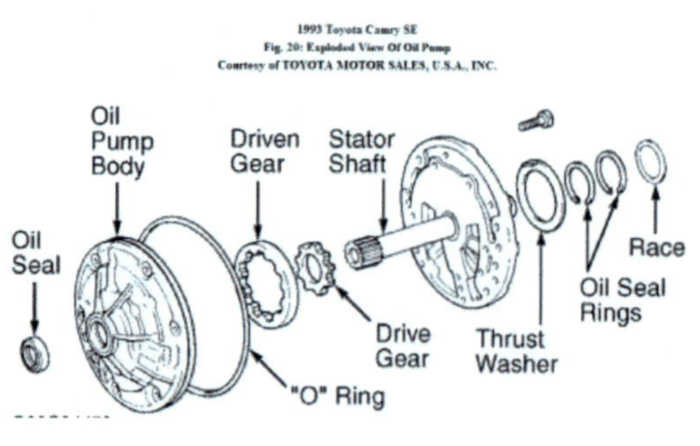

Oil Pump

The pressurized fluid is used to move valves in the
valve body, and to apply clutch packs and bands.
It is also used to lubricate the internal components
in the transmission, as well as carry fluid to the
transmission cooler

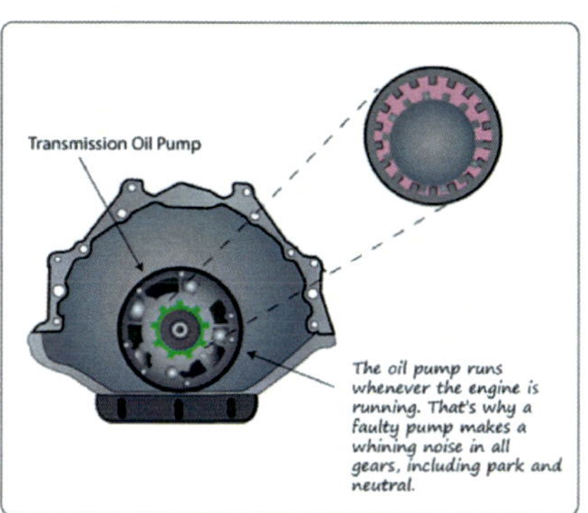

Internal Oil Circulation

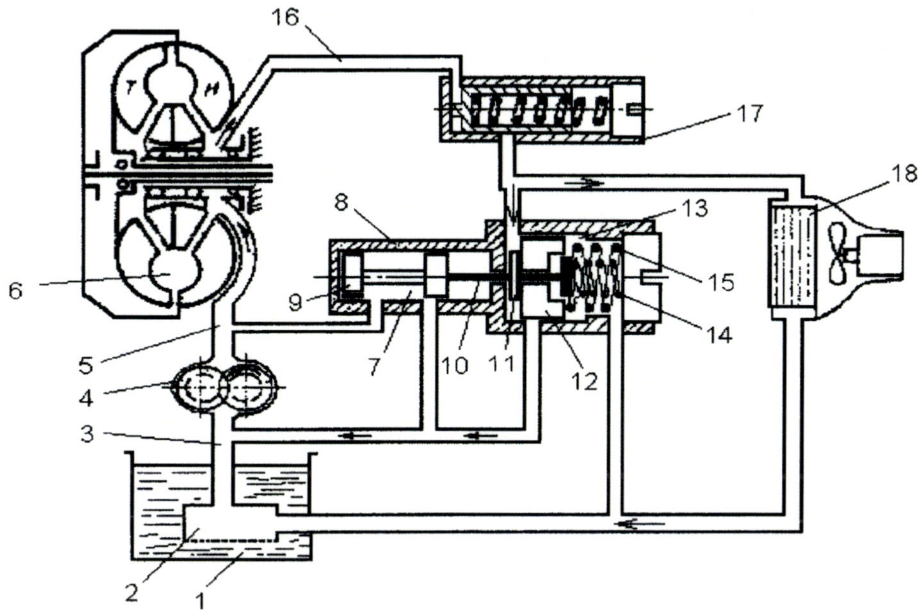

Oil Pump Assembly

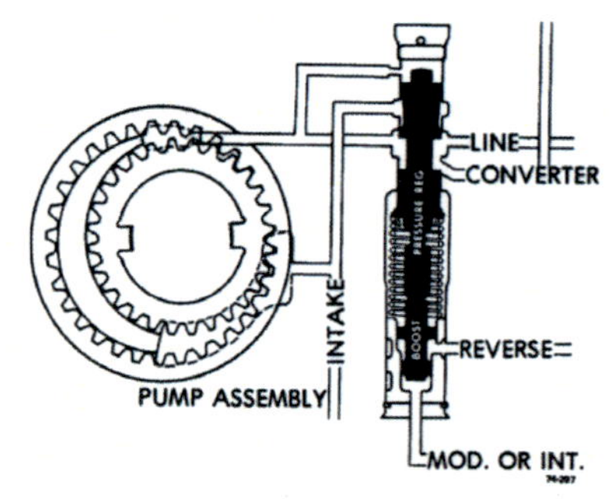

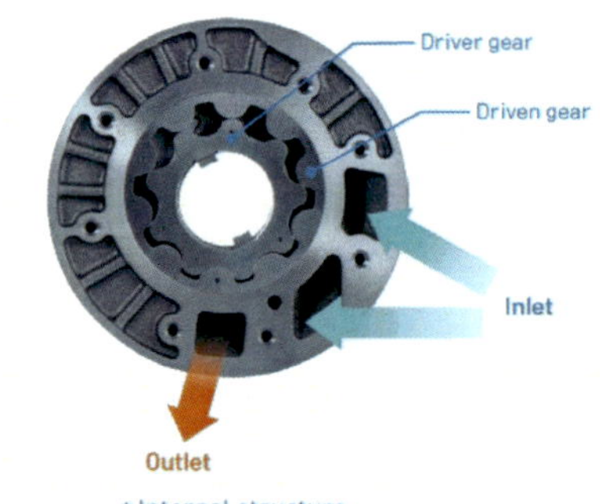

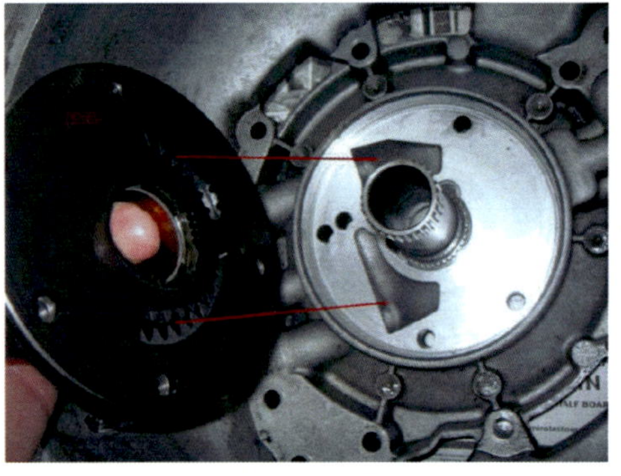

Oil Pump Noise

Some front pumps make a whining noise when they start to fail. Since the pump is directly connected to the motor via the torque converter housing, the motor must be turning for the pump to operate correctly

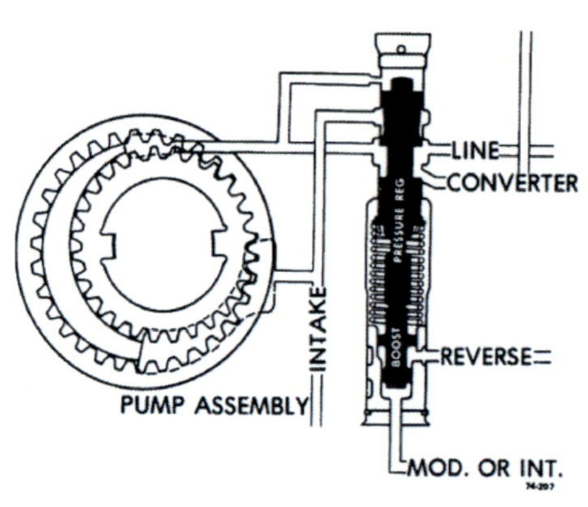

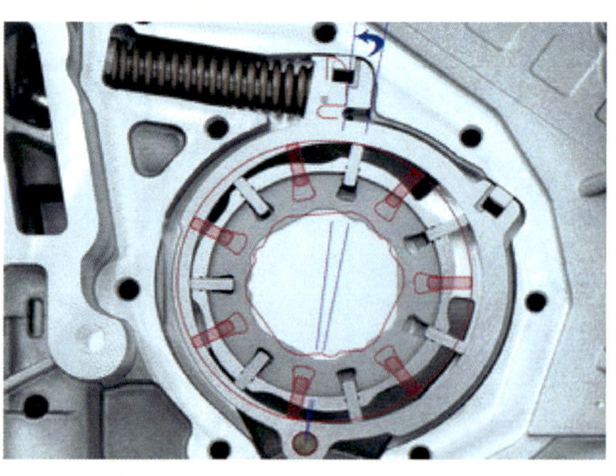

Oil Pump Pressure Adjustment

The best way to test a pump is to use a pressure gauge and attach it to the pressure port on the transmission. The minimum and maximum pressures are listed in the dealer's manual

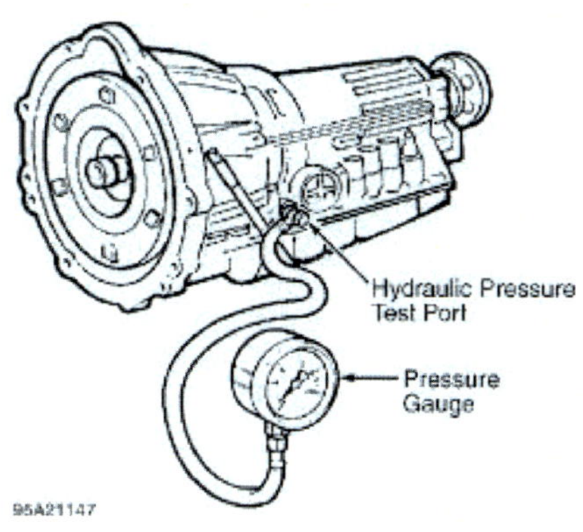

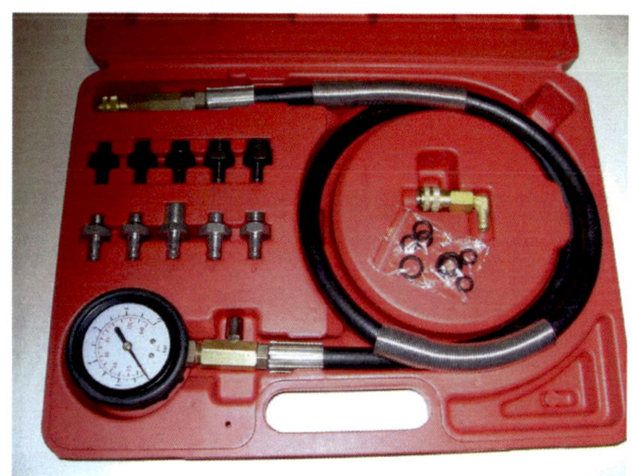

Oil Pump Problem

- Common problems with the transmission pump include:
    - Using gaskets of the wrong thickness
    - Broken drive (inner) gear
    - Worn tolerances between the inner and outer gear teeth
    - Worn tolerances at the front and back of the pump
    - A broken pressure relief spring, worn out or installed improperly

Valve Body

The valve body is the control center of the automatic transmission. It contains a maze of channels and passages that direct hydraulic fluid to the numerous valves which then activate the appropriate clutch pack or band servo to smoothly shift to the appropriate gear for each driving situation.

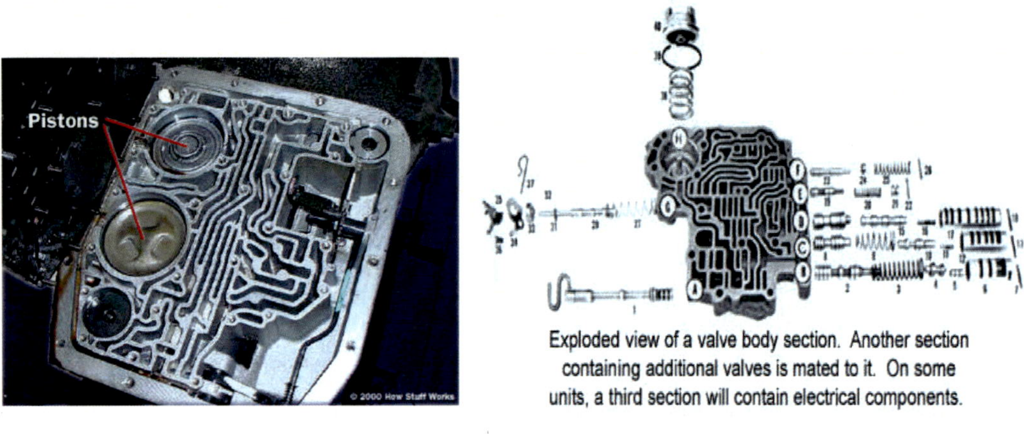

Exploded view of a valve body section. Another section containing additional valves is mated to it. On some units, a third section will contain electrical components.

Valve Body

The valve body is the control center of the automatic transmission. It contains a maze of channels and passages that direct hydraulic fluid to the numerous valves which then activate the appropriate clutch pack or band servo to smoothly shift to the appropriate gear for each driving situation.

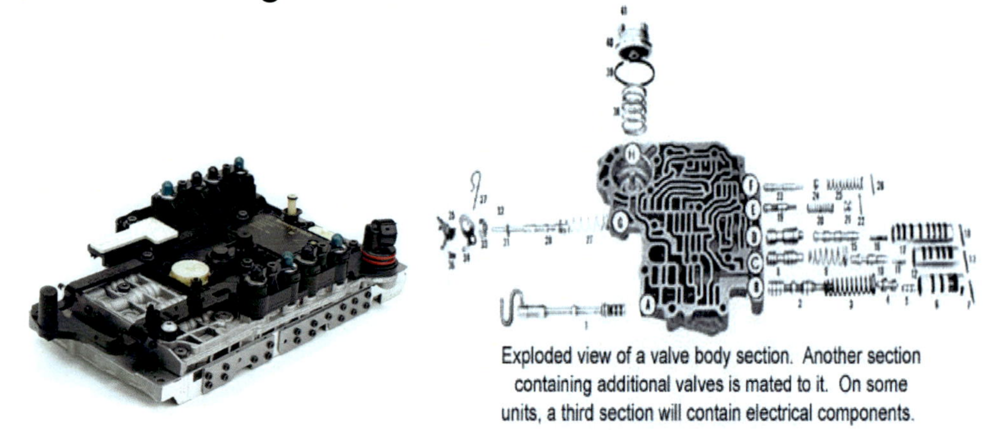

Exploded view of a valve body section. Another section containing additional valves is mated to it. On some units, a third section will contain electrical components.

Valve Body

The valve body is made up of tiny veins through which oil gets pushed by solenoids, thereby activating and deactivating the relevant clutch packs (gears)

Gain Adjustment

As in the generators. For a quick response during the acceleration . A set of high performance valves bolted at the body valve can be calibrated according with the manufacturer specifications

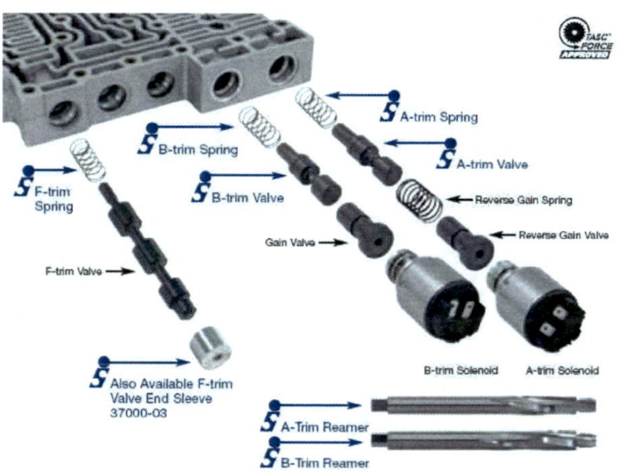

Hydraulic Schematic

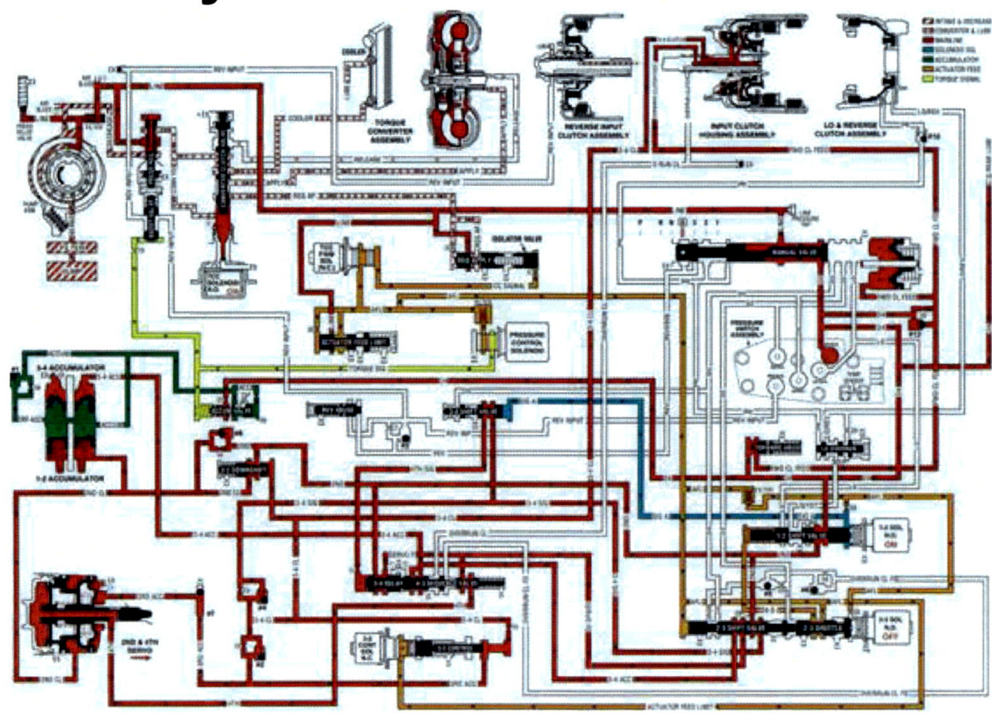

Hydraulic System

The Hydraulic system is a complex maze of passages and tubes that sends transmission fluid under pressure to all parts of the transmission and torque converter

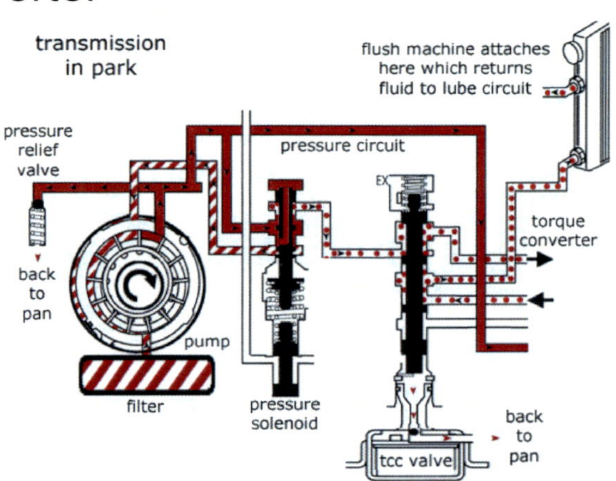

Hydraulic System Functions

Transmission fluid serves a number of purposes including: shift control, general lubrication and transmission cooling. Unlike the engine, which uses oil primarily for lubrication, every aspect of a transmission's functions are dependant on a constant supply of fluid under pressure

Hydraulic System Refrigeration

In order to keep the transmission at normal operating temperature, a portion of the fluid is sent through one of two steel tubes to a special chamber that is submerged in anti-freeze in the radiator .

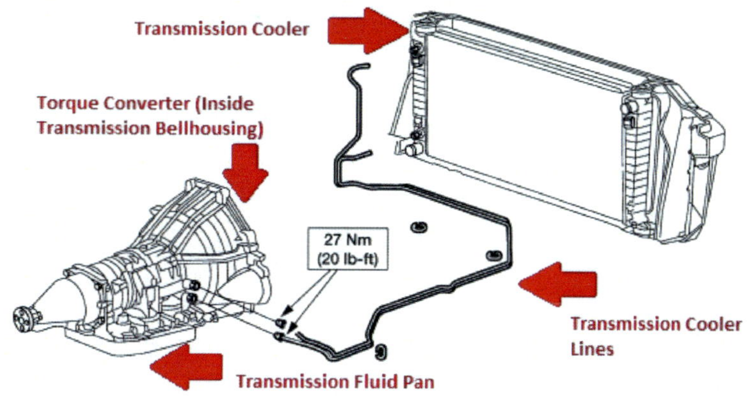

Transmission Oil Cooler

In order to keep the transmission at normal operating temperature, a portion of the fluid is sent through one of two steel tubes to a special chamber that is submerged in anti-freeze in the radiator (or Heat Exchanger). Fluid passing through this chamber is cooled and then returned to the transmission through the other steel tube.

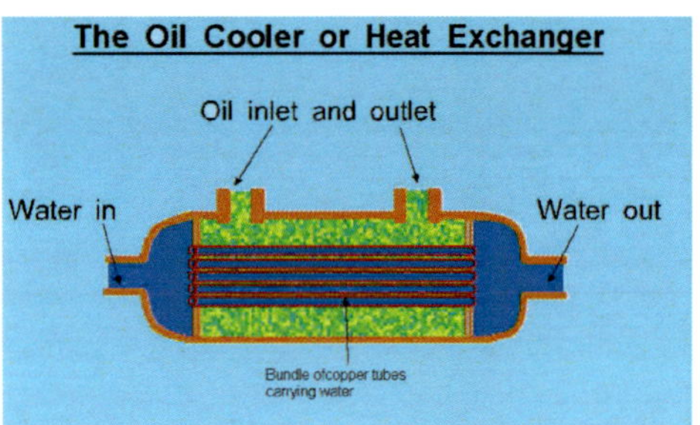

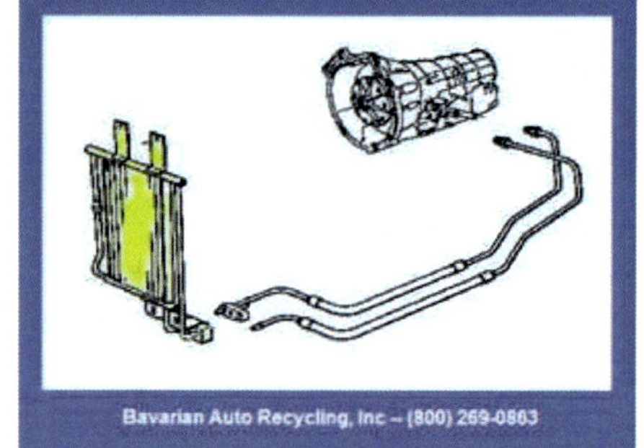

Transmission Oil Cooler

Constructed from a quality CU-NI (copper-nickel) alloy on the salt water side all furnaced brazed using silver alloy solders designed for "High-Pressure"

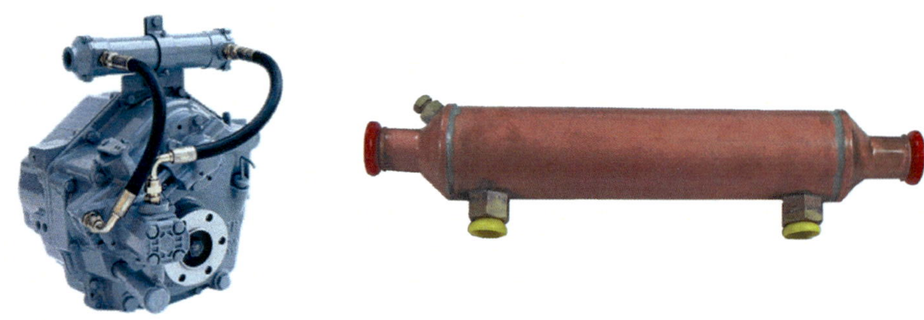

Hydraulic System

This is not unlike the human circulatory system (the fluid is even red) where even a few minutes of operation when there is a lack of pressure can be harmful or even fatal to the life of the transmission

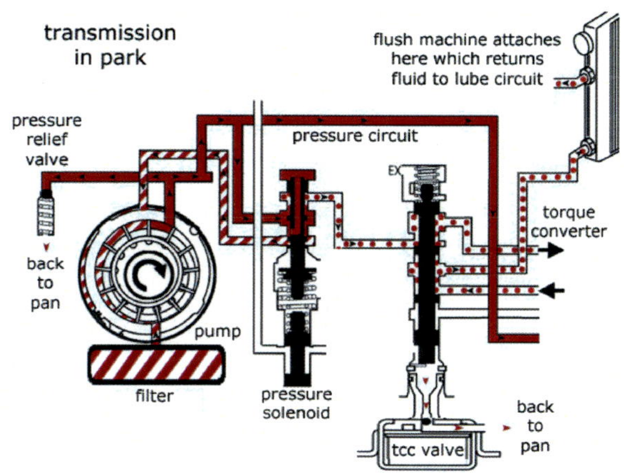

Transmission Oil Level

A typical transmission has an average of ten quarts of fluid between the transmission, torque converter, and cooler tank

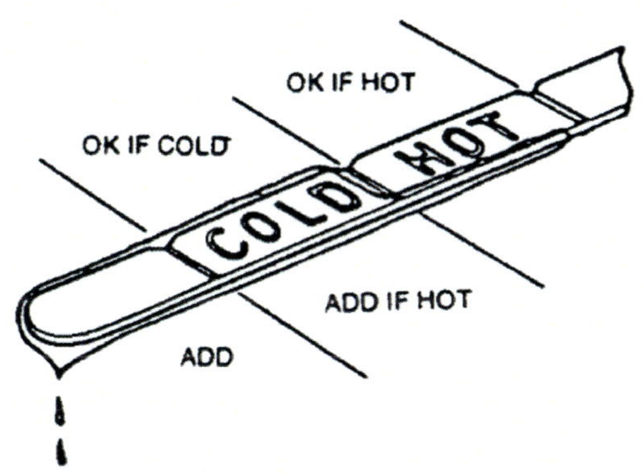

Transmission Oil Level

Most of the components of a transmission are constantly submerged in fluid including the clutch packs and bands. The friction surfaces on these parts are designed to operate properly only when they are submerged in oil

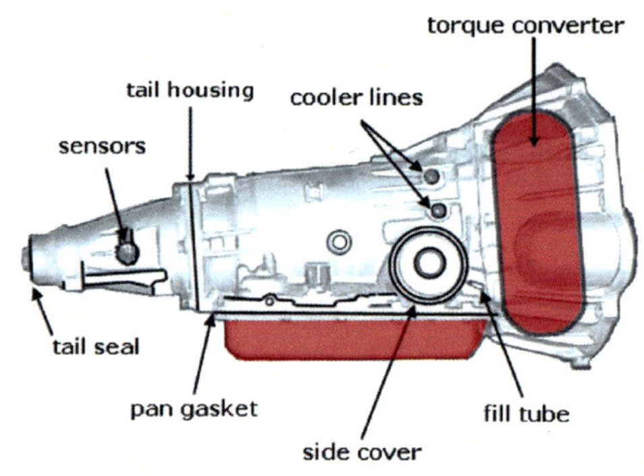

Transmission Service

In automatic transmissions, the recommended service interval is about every 30,000 miles or 30 months. (Check your owner's manual or service manual for your car's specifics.) The automatic transmission fluid (ATF) should be changed sooner if its dipstick reveals dark or burnt-smelling fluid.

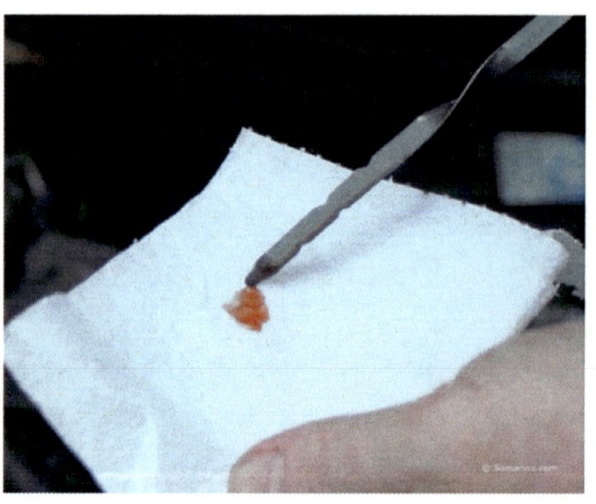

Transmission Fluid Service

This is the procedure to follow in order to keep the transmission in good conditions

Step 1: Transmission fluid drains better at operating temperature, so let your car idle for a few minutes first. After turning your ignition off, raise and secure the vehicle

Transmission Fluid Service

Step 2:Next, remove the bolts from one side of the transmission pan, being cautious of hot exhaust parts and fluid Gradually loosen the other bolts, which should allow the pan to tilt and begin to drain. Once all bolts are removed, lower the pan and dump the remaining fluid into the catch pan. Gently break the gasket seal with a screwdriver, if necessary

Transmission Fluid Service

Step 3: Clean the gasket surfaces on both the pan and the transmission housing. Inspect the pan for metal shavings or other signs of internal damage, and then clean it with solvent

Transmission Fluid Service

Step 4: Remove the old transmission filter and O-ring. The filter contains fluid, so keep the drain pan underneath

Transmission Fluid Service

Step 5: Install the new transmission filter, making sure that its O-ring seats in the appropriate orifice

Transmission Fluid Service

Step 6: Attach the new gasket to the pan **with oil-soluble grease** – not gasket sealer or adhesive or silicone is recommended

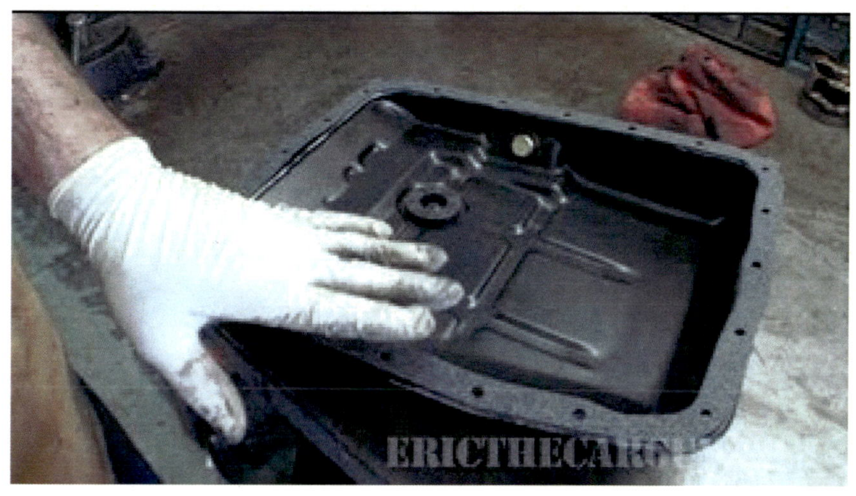

Transmission Fluid Service

Step 7: Refer to the service manual about using thread sealer on any or all of the transmission pan bolts, then screw in all fasteners finger-tight

Step 8: Torque the pan bolts to spec in a spiral pattern starting at the center. Maximum torque is often about 15 pounds per foot

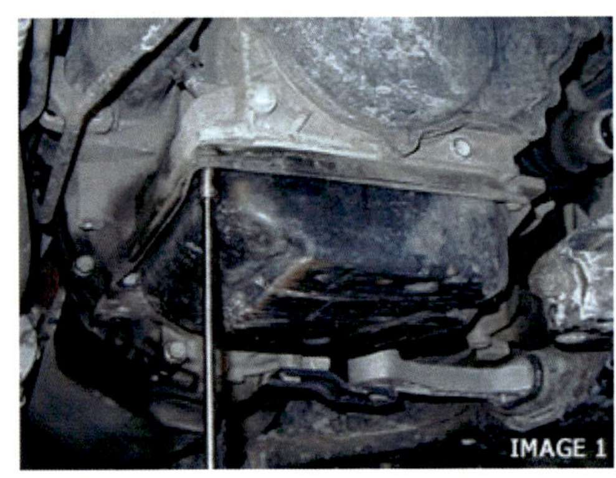

Chapter 5
Inboard Transmissions

Inboard Transmissions

The major difference between the marine diesel transmissions is the fact that the vast majority of marine gears have only one set of gears and therefore one ratio. (ZF makes a two-speed marine gear.)

Gasoline Engine

Diesel Engine

Forward & Reverse

Strictly speaking there is no such thing as forward and reverse, just clockwise and counterclockwise rotation. In some installations counter-rotation of the propellers is actually accomplished by simply swapping the linkage so that the opposite gear set is engaged when the lever is moved

In-board Marine Transmission

- An inboard transmission is a combination of a manual transmission with an automatic transmission
- In a marine transmission clutch or torque converter is not used . Due to this ,the engage of gears forward and reverse is abrupt and hard

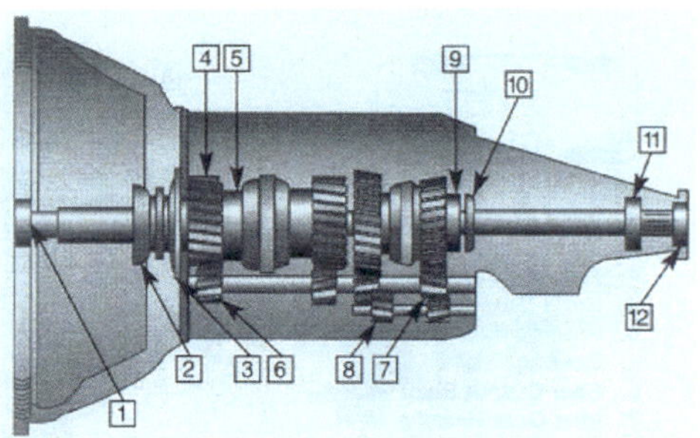

In-board Transmissions

- transmissions consist of a planetary gear set and multiple disc clutches.
- The input and output shafts are in line o V-Drive

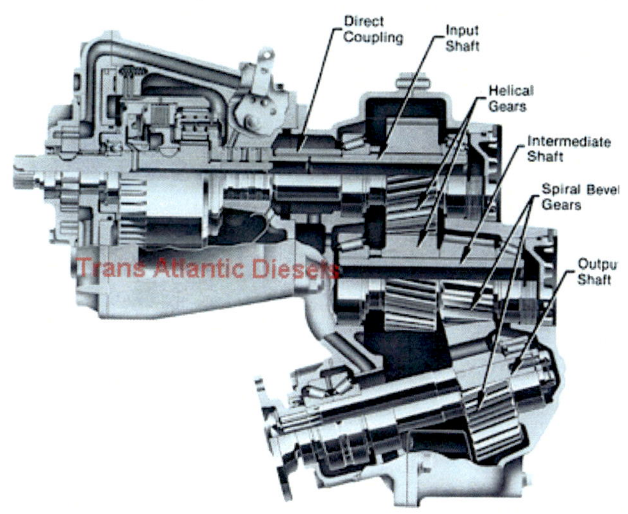

In-board Marine Transmission

- Most of the inboard transmissions use a unique gear ratio for forward and reverse
- The set of planetary gears just change the sense of rotation by means of a clutch pack and a couple of servo valves

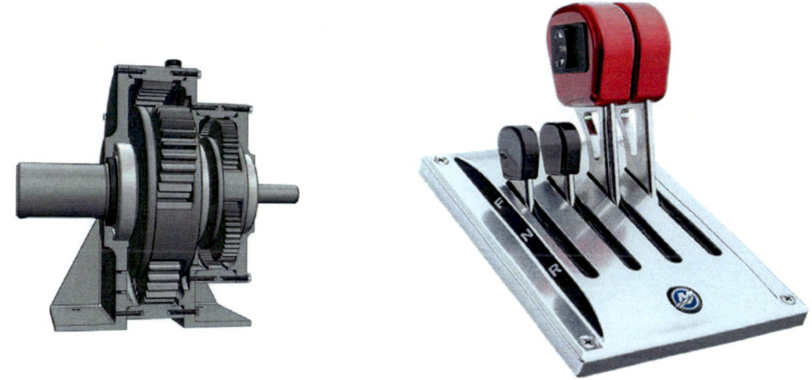

Plate Specification

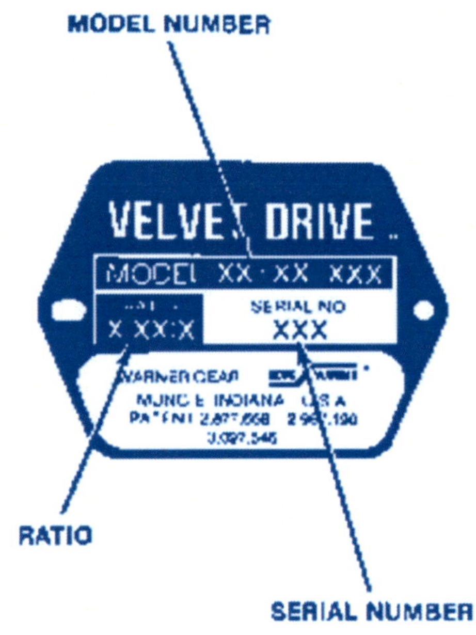

Drive Coupler Balancer

- The drive coupler balancer has two main functions
  - To absorb the excess of torque during the gear selection
  - To absorb the excess of vibration during normal operation

Drive Coupler Balancer

- A damper plate is bolted to the engine flywheel.

- The damper plate is splined to connect to the input shaft, The damper plate reduces torsional vibrations to the transmission from the engine

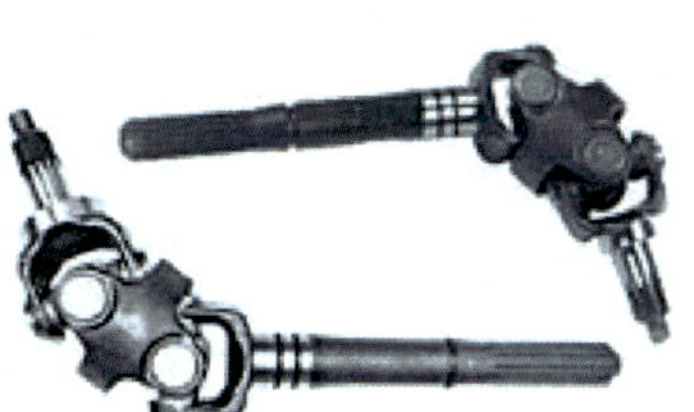

Drive Coupler Balancer

I strongly recommend that you change the damper plate whenever you have the transmission removed for service for several reasons. The "Always Change the Damper Plate" rule is because the plate is normally inaccessible, probably is rusted and frozen on the flywheel

Oil Pump

- The transmission oil pump is driven by the input shaft
- **It** supplies oil pressure to operate the clutch packs, lubricate parts, and provide cooling.

Oil Pump (Gas Transmission)

- Hydraulic pressure is provided by a crescent pump. Oil from the pump is sent to the control valve .
- The positions on the control valve are Neutral, Forward and Reverse

Oil Pump

Oil Pump (Diesel Transmission)

The oil pump is located in front of the transmission and receive the fluid from the oil cooler by means of hydraulic hoses

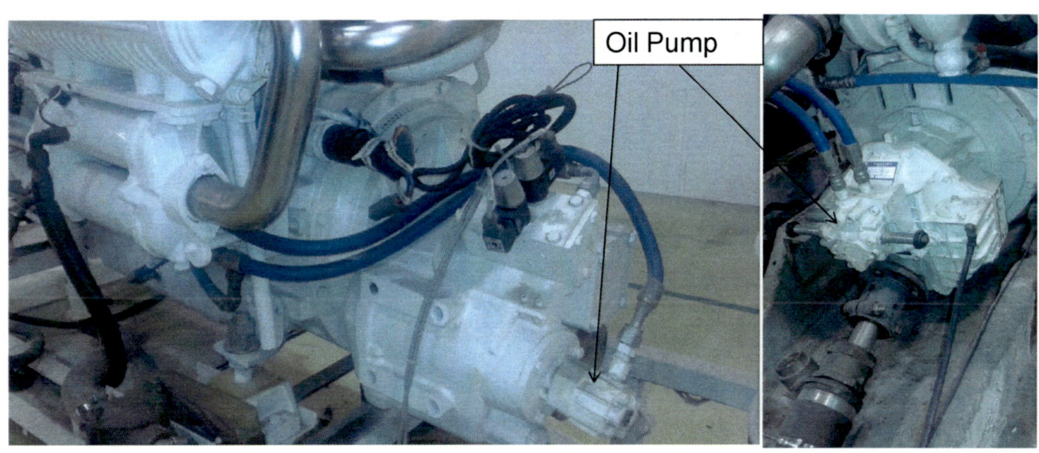

ZF Transmission Servo Valve

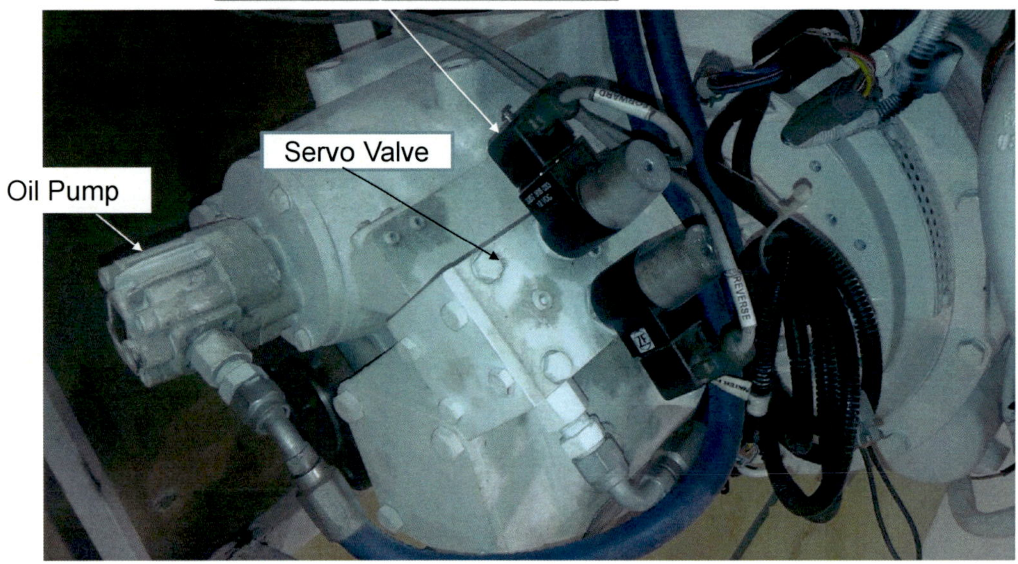

Control Valve

- The selector valve shifts the transmission from neutral to forward or reverse
- When selector valve is placed in the forward position , oil is directed to the forward clutch
- When the selector valve is placed in reverse position, oil is fed to the reverse clutch

Transmission Gear Ratio

Other function of the marine transmission is to set the ratio between engine rpm and propeller rpm. A typical diesel rotates at around 2000 rpm at cruise setting. A propeller turning this fast would be highly inefficient, so it is the job of the gears inside the transmission to slow down the propeller rotation to roughly half that of the engine, which is why this is always referred to as the reduction ratio

Gear Ratio

The Gear Ratio is fixed; the only way to change the ratio between engine and propeller rpm is by changing marine transmissions or changing the gear sets within the transmissions

Velvet Drive Gear Ratios

Forward	Reverse
1.0 : 1.0	1.0 : 1.0
1.52 : 1.0	1.1 : 1.0
1.88 : 1.0	1.68 : 1.0
2.10 : 1.0	2.07 : 1.0
2.57 : 1.0	2.83 : 1.0
2.91 : 1.0	3.20 : 1.0

Forward

The forward clutch is applied hydraulically when the shift lever is placed in the forward position.

Forward

This connects the input shaft to the output shaft. The unit then transmits power at a 1 to 1 speed ratio in the same direction of rotation as the engine

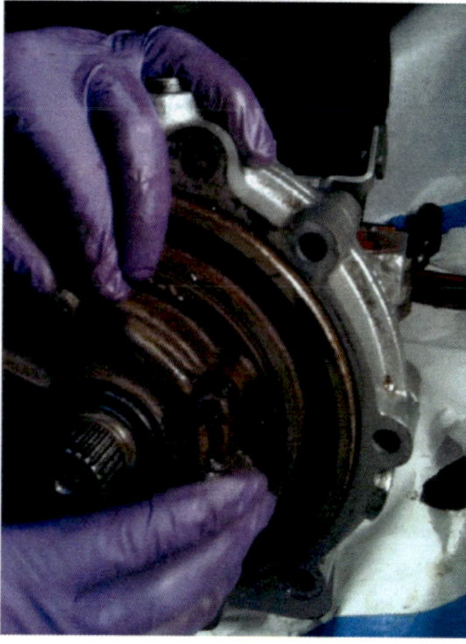

Reverse

The reverse clutch is applied hydraulically when the shift lever is placed in the reverse position

Reverse

The applied clutch holds the ring gear. The input shaft and sun gear , driven by the engine, drive pinions, which drive the carrier output shaft. The output shaft turns opposite to engine rotation at a 1,1 to 1 speed reduction ratio

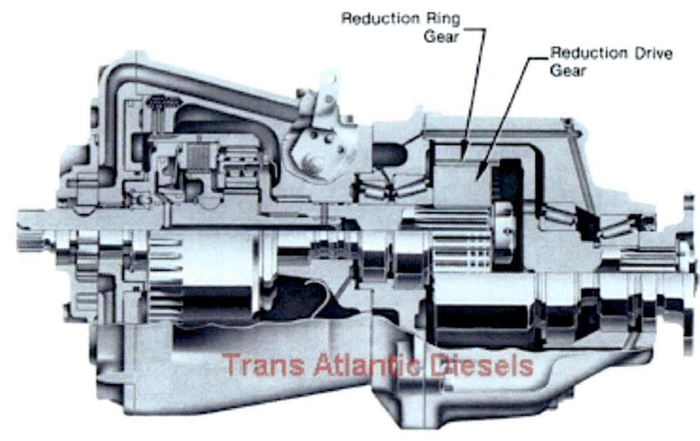

Forward & Reverse Solenoids

(Diesel Transmission)

Each clutch pack (for forward and reverse gear set) is activated by a solenoid that is powered from the shift control at the helm

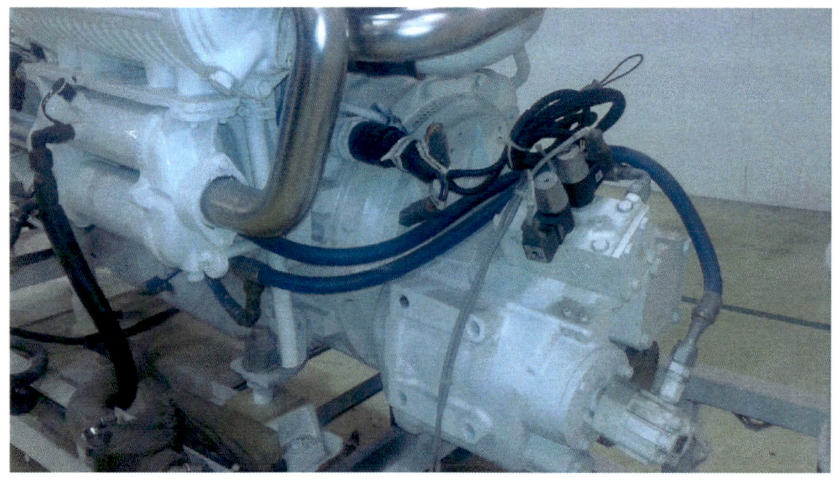

Oil Cooler

Each marine transmission has an internal oil pump that generates hydraulic pressure, which forces these clutches together to provide engagement. Because the pump generates heat as it pressurizes the oil, every marine transmission also must have an oil cooler

The oil cooler works at operating pressures of 300-500 PSI even if the transmission has lower operating pressures

Transmission Oil Cooler

Due to the high loads applied the transmission oil should be cooled by means of a hydraulic pump and a heat exchanger

Oil Cooler

Each marine transmission has an internal oil pump that generates hydraulic pressure, which forces these clutches together to provide engagement. Because the pump generates heat as it pressurizes the oil, every marine transmission also must have an oil cooler

The oil cooler works at operating pressures of 300-500 PSI even if the transmission has lower operating pressures

Type of Oil (Gasoline)

Dexron, Type F, or any hydraulic fluid which meets the C-3 oil specification is acceptable. Do not mix different brands

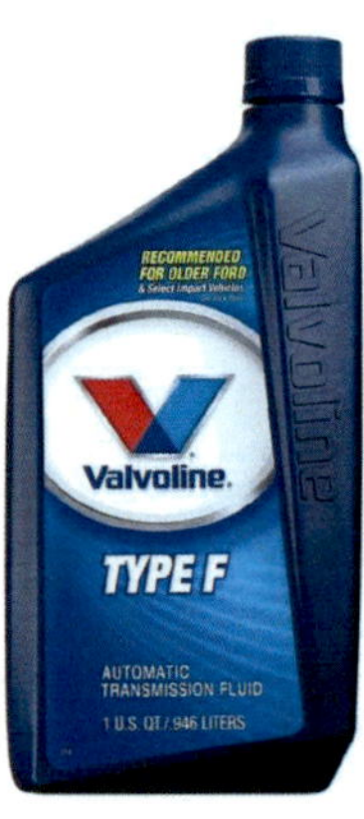

Type of Oil (Diesel)

The ZF/Hurth Workshop Manual indicates that automatic transmission fluid or ATF is to be used in these mechanical gears. ATF, commonly known as Dextron II or III, is far preferable to the thicker SAE 90 hydraulic gear oil

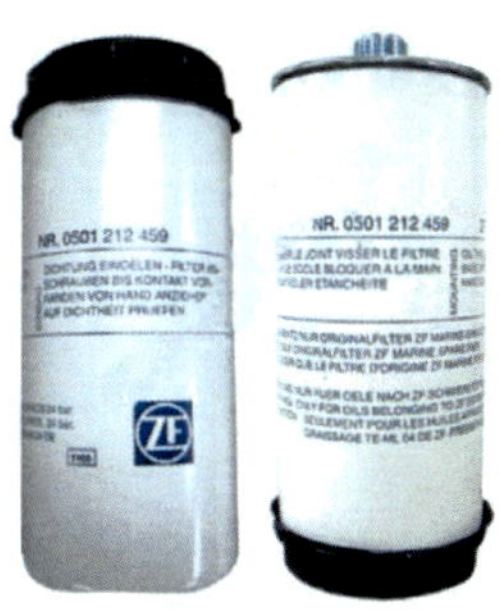

Oil Temperature

- If the transmission oil temperature has exceeded 190 F or the alarm sounds, the oil must be changed in the transmission and cooler system.

- If engine doesn't exceed 3,000 R.P.M., a Dextron II oil is acceptable. SAE #40 and multiviscosity oils are not recommended.

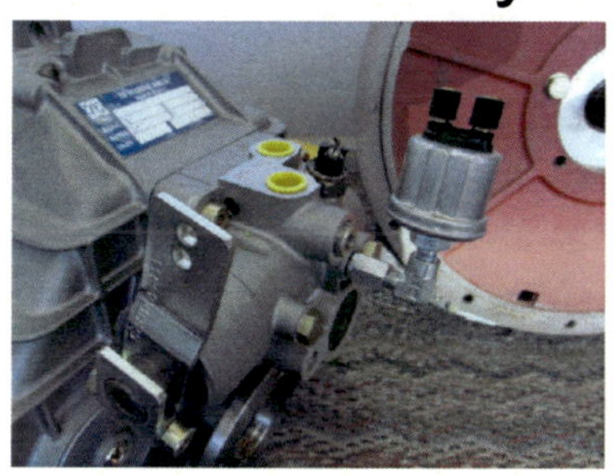

Oil Change Cycle

Fluid in Hurth marine gears should be changed at the start of each season and at 250 hour intervals. Change it more often, perhaps every 100 hours, if you are using the marine transmission in a severe duty cycle such as launch duty or the placement of moorings or towing

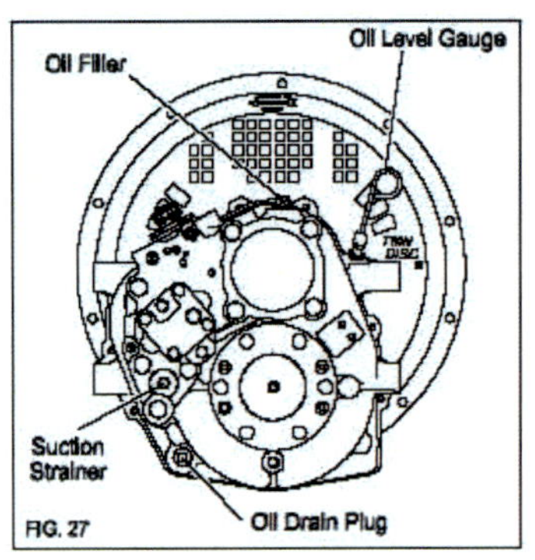

Oil Change Cycle

Change it immediately if you have tangled the prop in lobster pot wrap, overheated the gear, or if the fluid is discolored

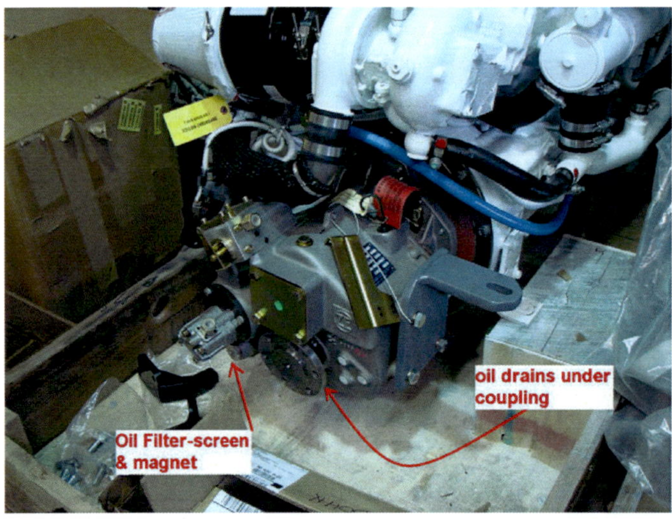

Velvet Transmission Repair

The front pump housing shall be marked. The pump can be mounted for left and right rotation so its is crucial mark it

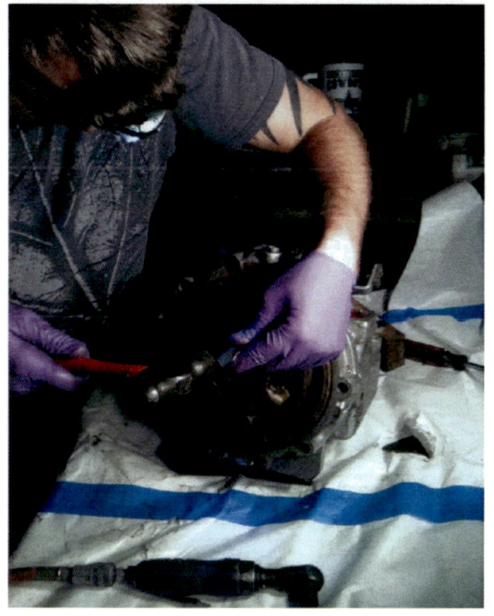

Velvet Transmission Repair

Remove the front cover pump

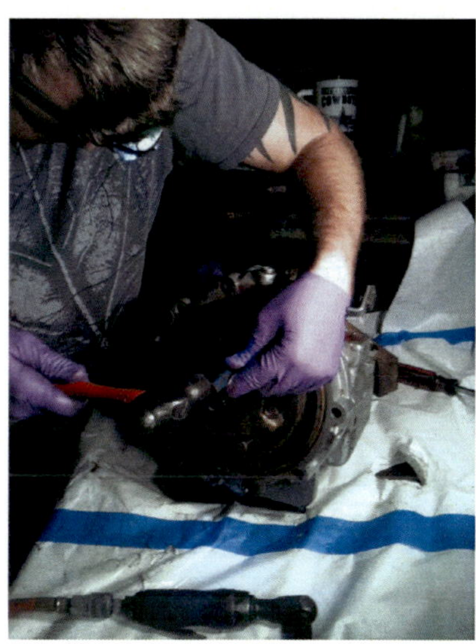

Velvet Transmission Repair

Remove the pump cover stator. In hydraulic terms this is part of the pump. The symptom when the pump is working bad is an slow response of the transmission when you apply forward or reverse

Velvet Transmission Repair

Remove Pump Housing with small pick or (tweaky flathead), clean shaft from seal pitting and debris for pump gear to be removed

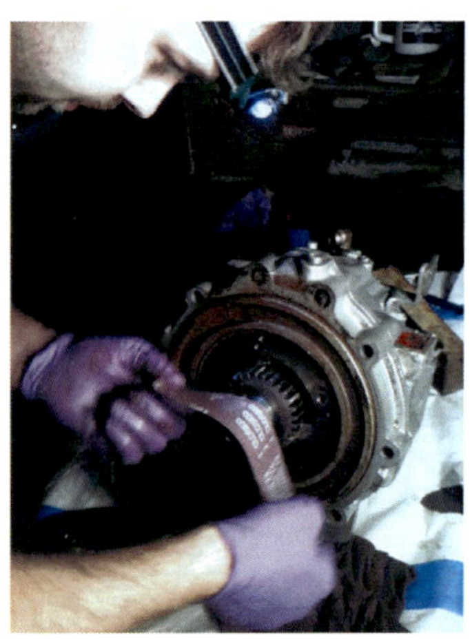

Velvet Transmission Repair

Remove The crescent pump gears

remove pump drive gear and remove woordruff key from drive gear assembly

Velvet Transmission Repair

Remove case bolts (12 points (10mm) 3/8" socket necessary)

Forward & Reverse Adapter

Remove forward and reverse adapter
Note sealing ring is sitting on adapter

Pressure Plate

Remove reverse clutch pressure plate ((it's sitting on
the adapter in picture), outer clutch plate and
reverse clutch assembly stage in order

Drive Gear Removal

Without any tool (just with your hands)pull the
drive gear assembly

Snap Rings Removal

Remove snap ring (this one is very thigh and often
gets warped with removal, that's why you might
want to order it with the rest of the parts) and the
second snap ring (use snap ring pliers)

Drive Gear Removal

Slide the drive gear out of the ring gear (I used the black mallet to help it out) (you'll be replacing the insides anyway)

Snap Ring Removal

Set drive gear aside and let's take look at the ring gear. Remove snap ring and remove the bearing

Forward Clutch Cylinder

Now the forward clutch cylinder should come out now. You can separate it from the clutch spring (note the direction of the spring , the lower side of the dish is closest to shaft)

Clutch Plates Inspection

Check the condition of clutch plates inside the ring gear now. In general i recommend replace the clutch plates

Clutch Plates Removal

Remove snap ring , clutch pressure plate and the clutch plates (stage to remember order when new ones go in). Take pictures of the sequence

Valve Spring Assembly

Clean the transmission case extremely and then remove the valve spring assembly

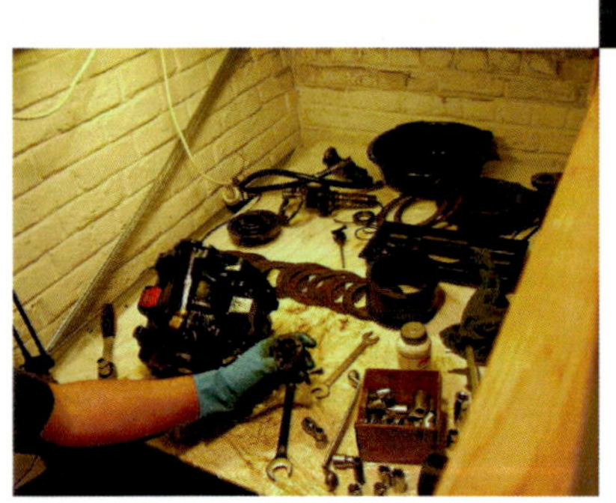

Shaft Nut Removal

With an impact gun and using a 1-1/2"
(38mm) socket , remove the main shaft nut

Coupling & Bearing Removal

Now remove the coupling and the bearing

Pinion Cage & Bearing

Remove pinion cage ; put the bearing in WD40 bath ; remove oil baffle and remove old gasket.

Oil Strainer

Remove and clean the oil strainer

New Plates Preparation

A bath into ATF oil is recommended for the new plates before the installation

Assembly Process

- After this point the transmission will be ready for the assembly process following the same procedure with enough amount of grease and ATF oil in each step
- Start with the clutch pack, adding oil between plates

Clutch Spring

Next comes clutch spring and forward clutch cylinder . The clutch spring is dished. the lower side of the dished spring should be closest to the shaft.

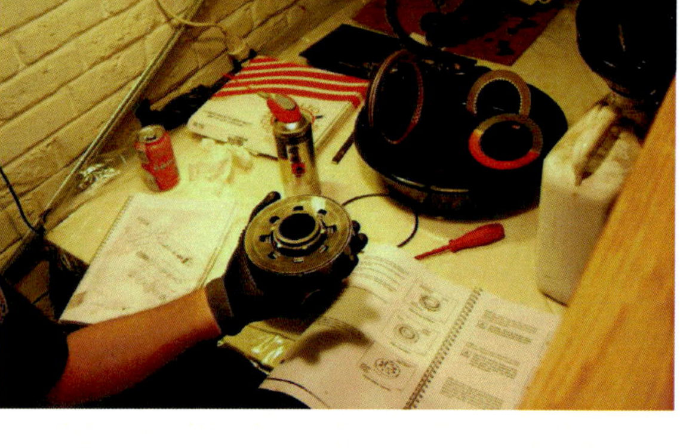

Snap Ring

Place spring and forward clutch cylinder in ring gear and place snap ring . **You will need a press at this point.** Pressing down on the forward clutch cylinder will allow to place snap ring in its groove

Clutch Pack

Your clutch pack and drive gear are now ready. You can now return the oil baffle to its original location in the case

Bearing Retainer

Next replace the oil seal and then placed it in bearing retainer

Main Bearing Case

Place bearing in case. the groove in the bearing goes towards the coupling (propeller) side. Your kit should contain an O-ring that goes into this groove

Bearing Cover & Gasket

Place gasket and bearing retainer . Use a torque wrench to place bolts. 27-32 lbs-ft I used ARP on bolts

Thrush Washer & Clutch Pack

Install thrust washer when putting in the
clutch pack and drive gear

Pressure Plate & Dowel Pins

Place pressure plate springs and greased
dowel pins

Reverse Clutch Plate

Place your new reverse clutch plate assembly and .
Bronze goes in first, then steel, then bronze again

Forward & Reverse Adapter

Place forward and reverse adapter . Bolt
with torque wrench 27-37 lbs-ft

Pump Housing

Place the drive gear, gasket and pump housing with a wrench to torque 17-22 lbs-ft. The pump housing has to be placed in the same position as when you disassembled

Pump Housing

Place the drive gear, gasket and pump housing with a wrench to torque 17-22 lbs-ft. The pump housing has to be placed in the same position as when you disassembled

Oil Strainer Assembly

Next comes the oil strainer assembly . Replace in same order as when you disassembled

Final Product

At this point the transmission is ready to be coupled to the engine

Shifting Problems?

First check and adjust your linkage. Linkage problems are very common on Hurth marine gears when mounted to auxiliary engines. If the problem is still there, see if there is a difference in the difficulty in engaging forward versus reverse

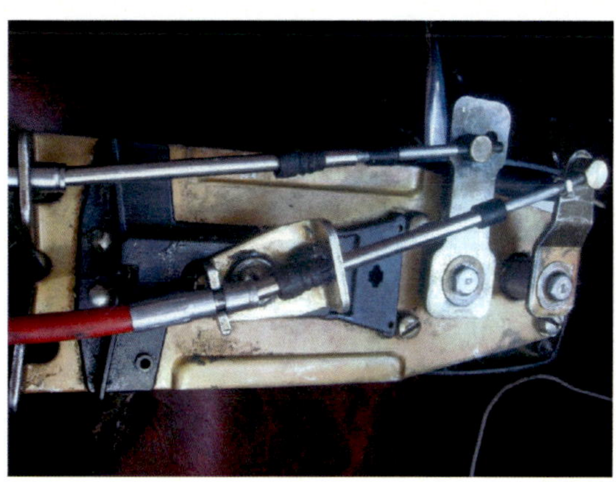

Shifting Problems?

- If it is significantly more difficult to engage the marine gear in forward as opposed to reverse, this indicates that the thrust washers are worn. The gear will require service or replacement.

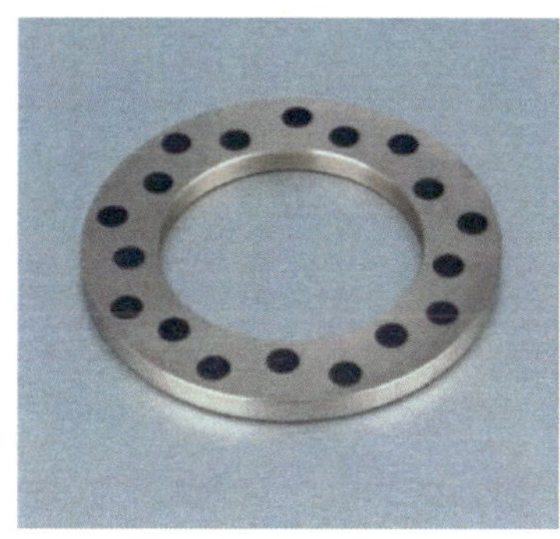

Slow Traction Response

This is a typical case of pump defective. the pump should be replaced. **Attention:** A slow traction response is different than a transmission with slipping

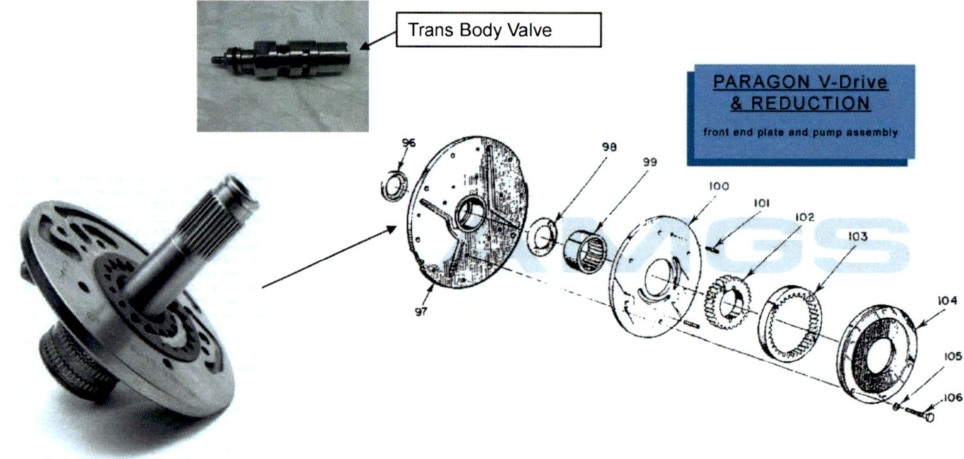

Slow Traction Response

A slow traction response is a similar concept than the "Gain" in electricity. It is basically the time that the transmission need to engage the gear once the shift control is applied. If this time is more than 1 second probably the pump and the body valves are defective

Transmission Slipping Symptoms

- One common sign of a transmission slip is when the revolutions per minute (RPM) gauge is above 3,000 (for Gas Engines) or above 2500 for Diesel Engines

- Another sign of a transmission slip is when the boat is taking too much time to advance with a surge at the engine rpm

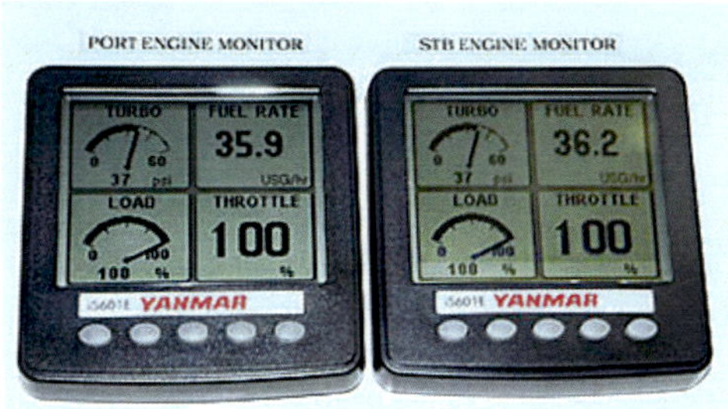

Transmission Slipping Symptoms

Another cause is when the friction material on the clutch pack worn out or broken. With this type of problem, you will not only have to deal with the transmission slipping but a multitude of engine troubles, including worn out transmission gears

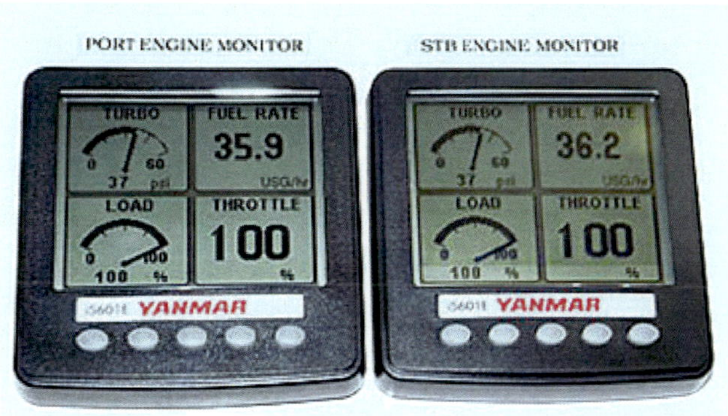

Chapter 6
Coupling, Shaft &
Propulsion System Alignment

Coupling and Shaft

Before proceeding with the alignment check that the diameter of the coupling in the transmission is equal to coupling at the shaft. Also check that the holes be aligned between both couplings

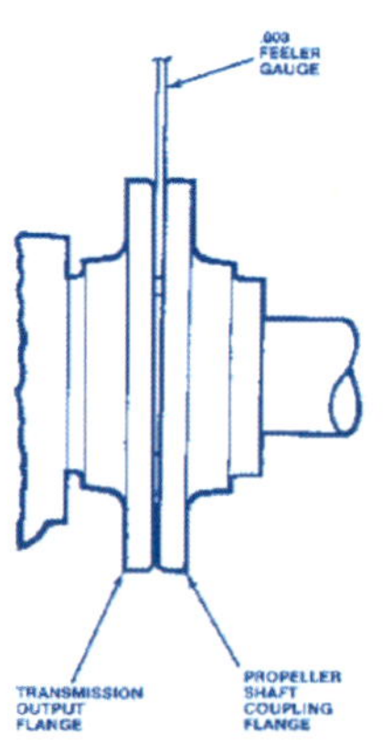

Coupling Vibration Absorber

If you suspect that the result of the alignment will not perfect . you can install a vibration absorber between both couplings to avoid excessive vibration

 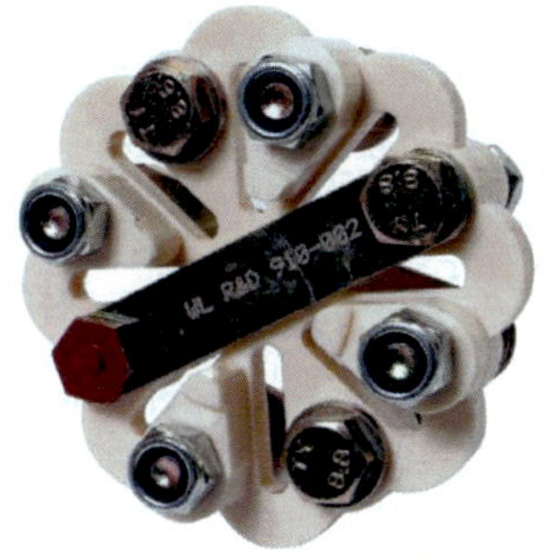 

Coupling Vibration Absorber

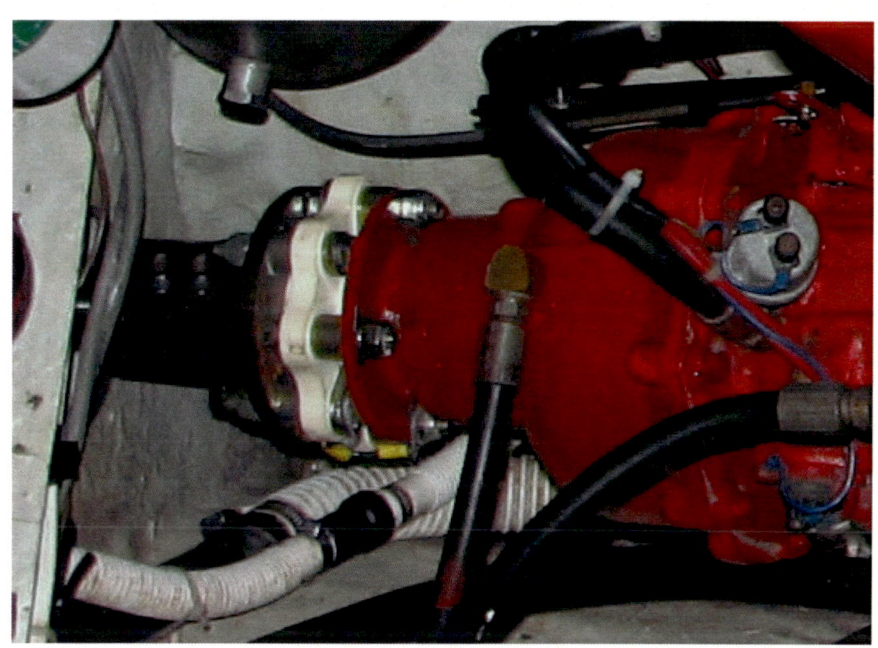

Propulsion System Alignment & Repair

A common problem occurs when the bottom of the boat hits a rock. In this case the entire propulsion system and hull must be repaired

Propulsion System Alignment

Usually when the propeller hits the ground, additionally the strut , shaft , rudder and boat hull also are affected

Boat Propulsion Alignment

The first step is remove the entire propulsion system ;next ,repair the hull ; then replace the affected parts and finally do the alignment of the system

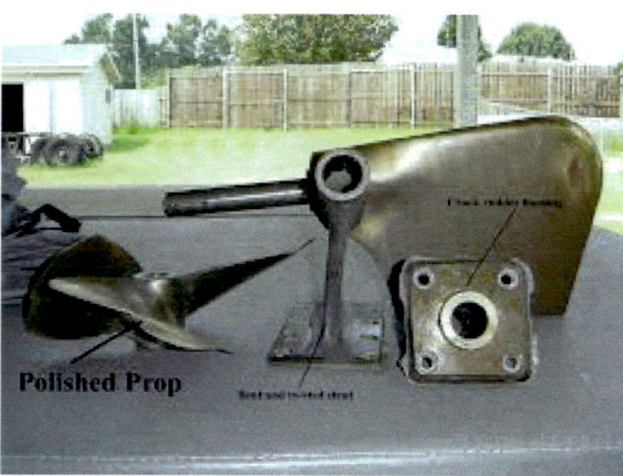

Propulsion System Alignment

- Components who are significantly affected should be replaced

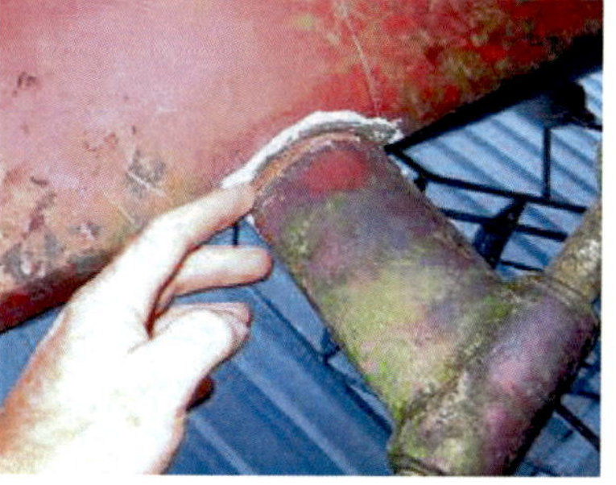

Propulsion System Alignment

- The fiberglass components must be repaired by an experienced professional
- Your basic knowledge in fiberglass repairs are not sufficient for this type of project

Propulsion System Alignment

Once repaired the hull and metal components, we can begin the process of alignment

First we will provisionally install the metal components (Flange, Strut and Rudder)

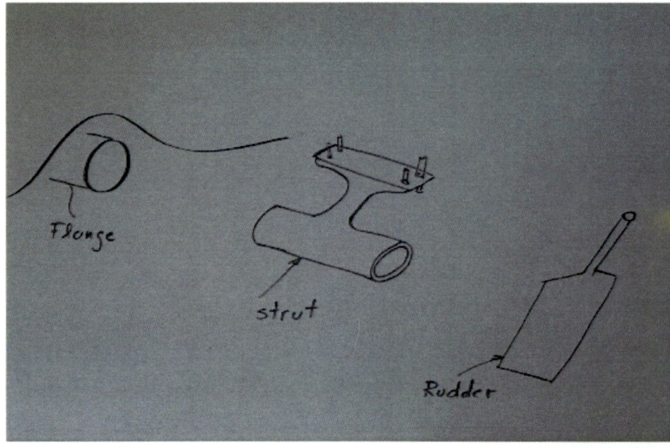

Boat Propulsion Alignment

- Pass a thin steel wire through these components
- Keep the wire stretched and centered by a heavy object on the other end

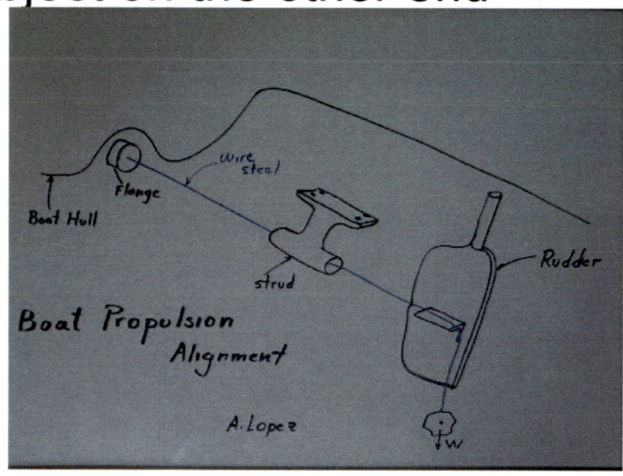

Propulsion System Alignment

By means of a compas verify that the wire is properly centered and aligned at both ends of the strut

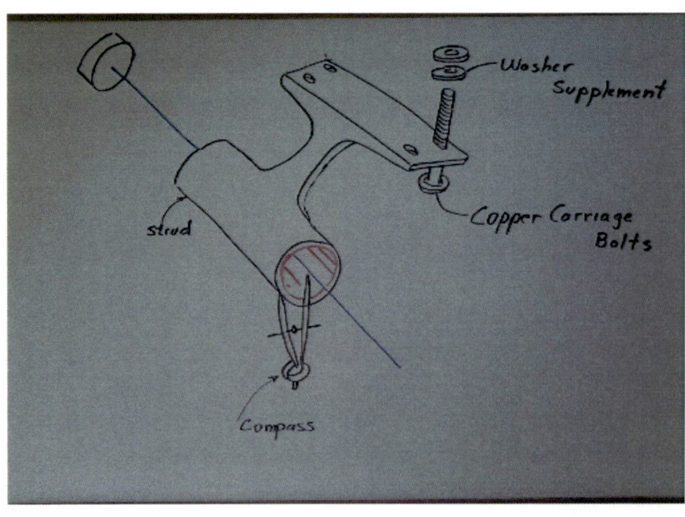

Boat Propulsion Alignment

- Adding or removing washers between the hull and the strut, the system can be aligned properly

- Do not add any sealant until the system is aligned

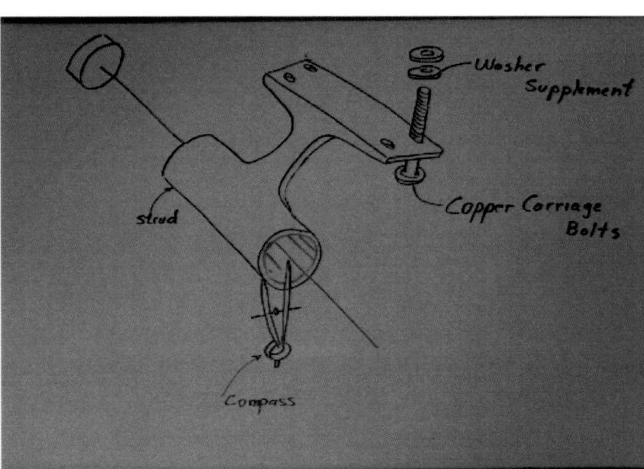

Boat Propulsion Alignment

- Once the system has been properly aligned, remove the bracket and add the marine sealant, considering the number of washers that must be installed on each bolt

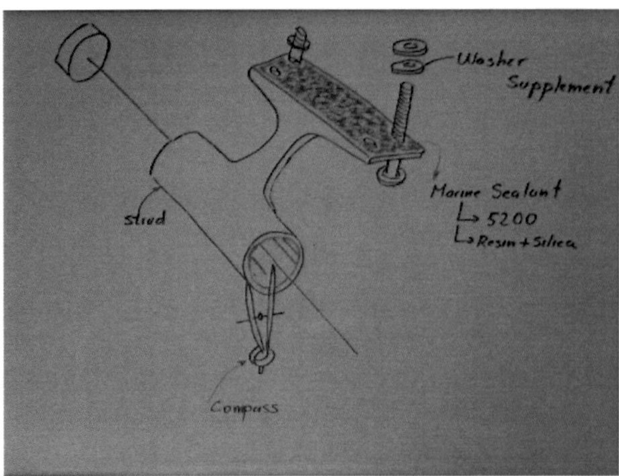

Boat Propulsion Alignment

Gently tighten each bolt until the surface is perfectly adjusted. Wipe excess sealant and let dry a few hours before installing the shaft

Shaft Installation

Insert the shaft spinning it with your hands , keeping it lubricated with soap or shampoo. Don't try to use oil or grease

Shaft Installation

The shaft should turn freely with your hands before installing the propeller

Propeller and Shaft lapping

Before adjusting the propeller make sure that the contact area between the shaft and the propeller is properly lapped

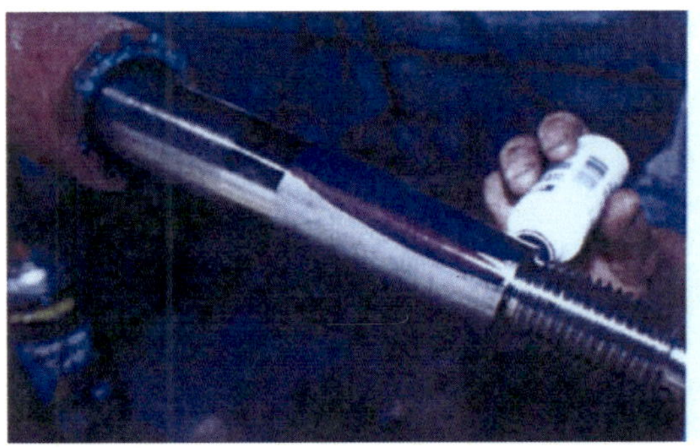

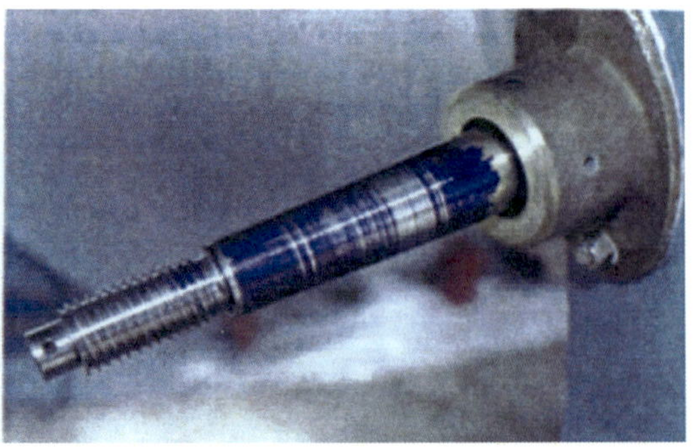

Nut and Cutter Pin

Tighten nuts and fasten it properly using either pin or safety wire

Shaft and Coupling

Insert the shaft into the coupling using a piece of wood to protect the face of the coupling

Shaft and Coupling

Fasten the coupling by means of the set screws and safety wire

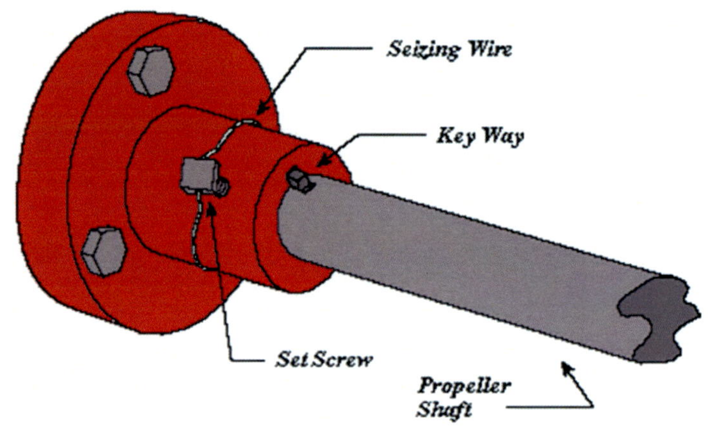

Compatibility Couplings

Verify that both the coupling of the shaft and the coupling of the transmission are mechanically compatible

Engine Alignment

By means of the motor mount adjustment both couplings can be aligned

Engine Motor- Mounts

- Motor mounts have likely been designed to minimize transfer of noise and vibration to the hull of the vessel .

- Using non-standard mounts in order to "improve" the mounting position of the engine may create more problems than it yields advantages

Shaft and Transmission Alignment

Analyze the offset misalignment between the axis of the propeller and shaft of the transmission

This deface can be corrected by adjusting nuts on each motor mount

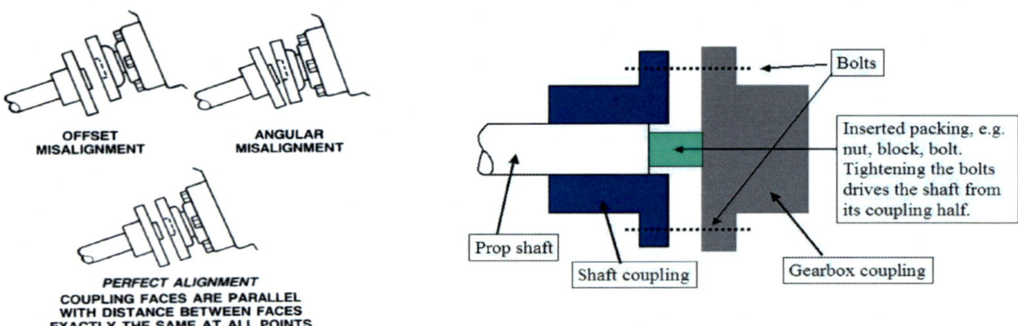

Parallels between the couplings

Once shafts and couplings have been aligned, check the parallelism of their faces using a feeler gage and turning the shaft slowly with your hands

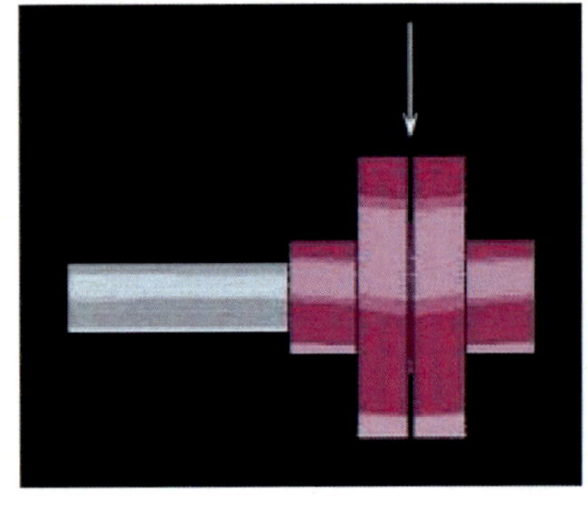

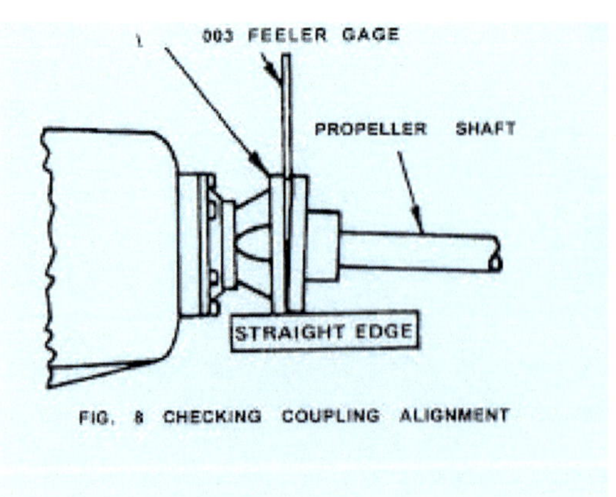

Shaft Couplings Alignment

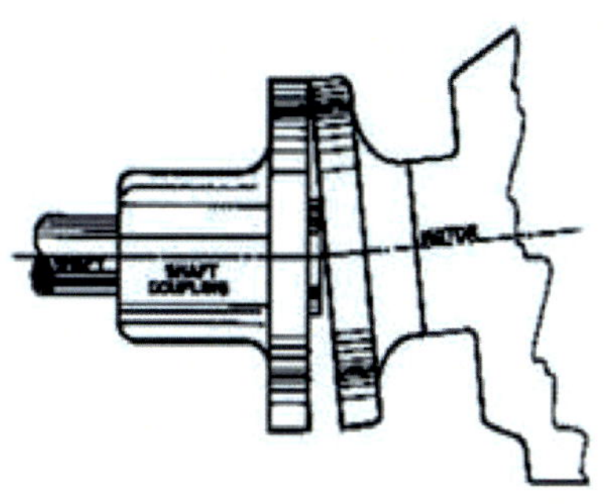

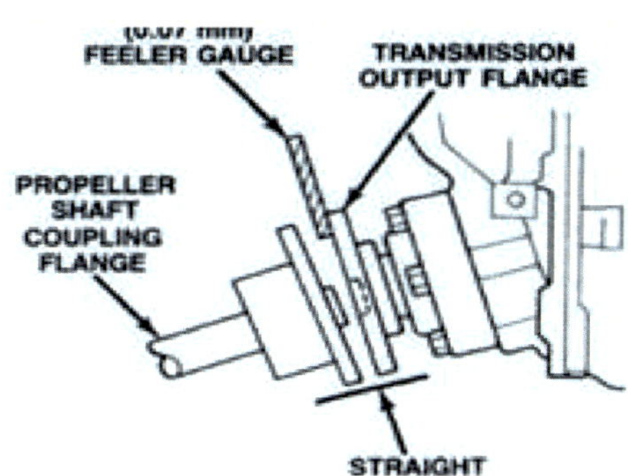

Shaft alignment

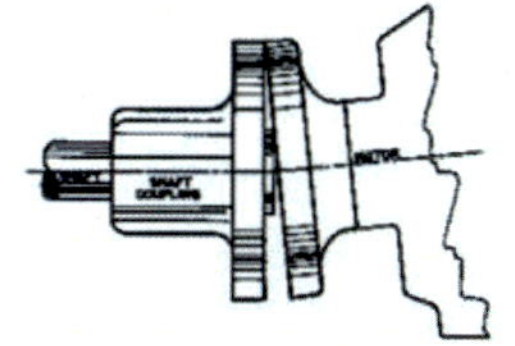

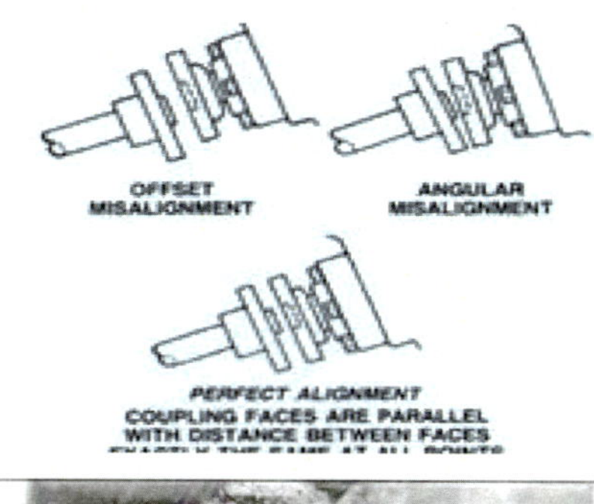

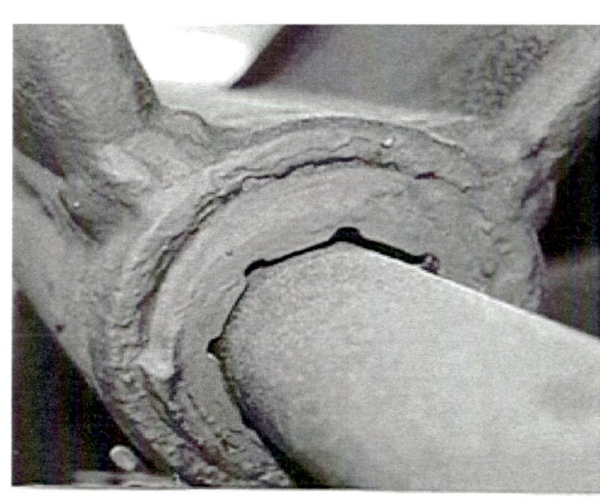

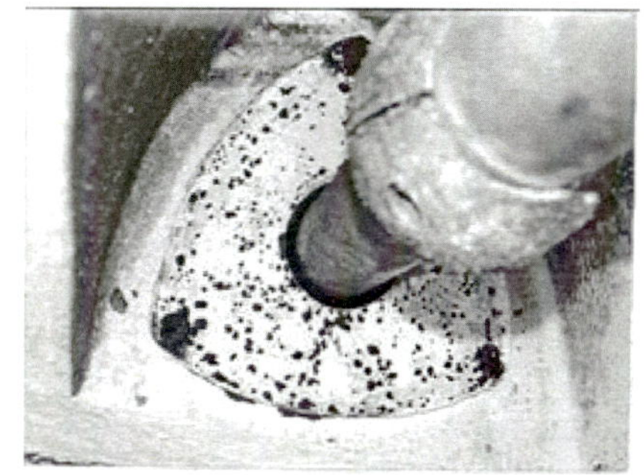

Checking Measures

Once the alignment process is finished verify that each element is installed within the range approved

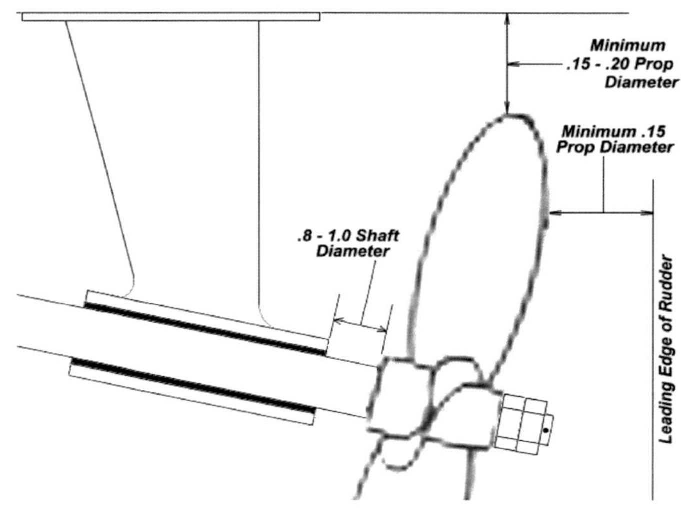

Shaft center Line Prop

Dave Gerr , recommends that for an inboard single engine boat. The maximum shaft angle should be about 3 degrees

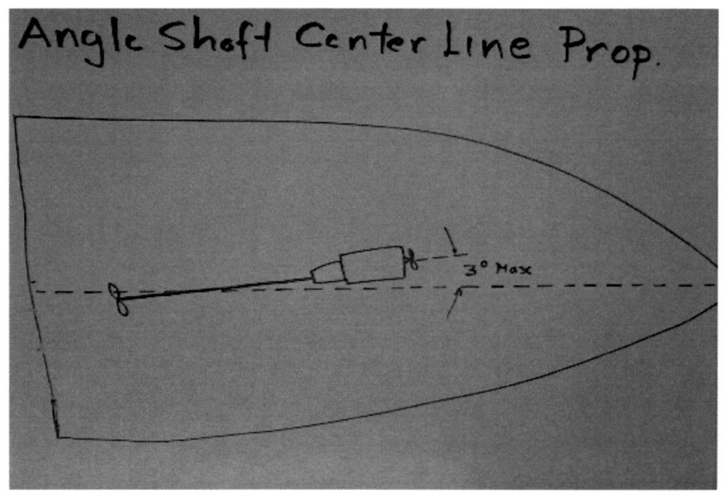

Offset Distance

The offset distance shouldn't be more than
about 8 percent of beam

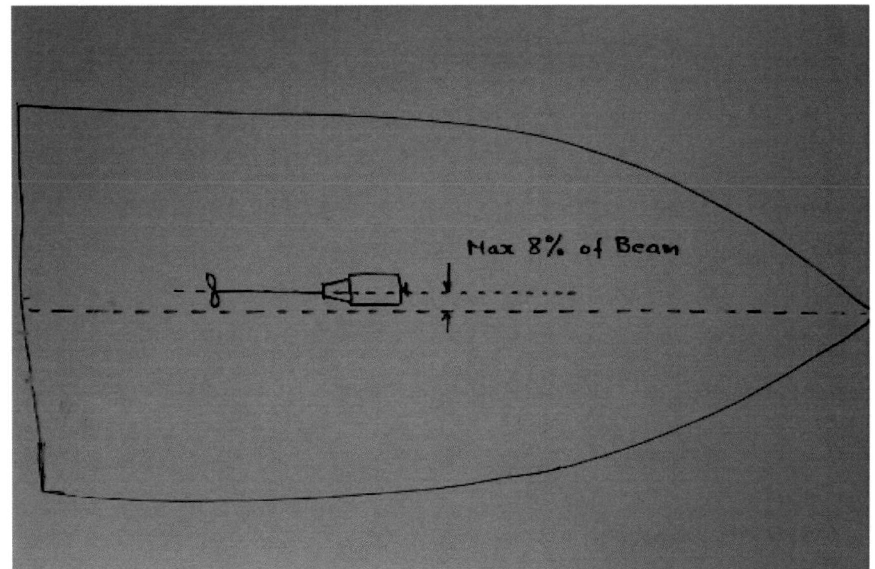

Twin-Screw Angled-out Shafts

- The Goal is to allow higher engines to fit lower in the bilge of the V of the hull body
- The max. Shaft angle should be 2 degrees

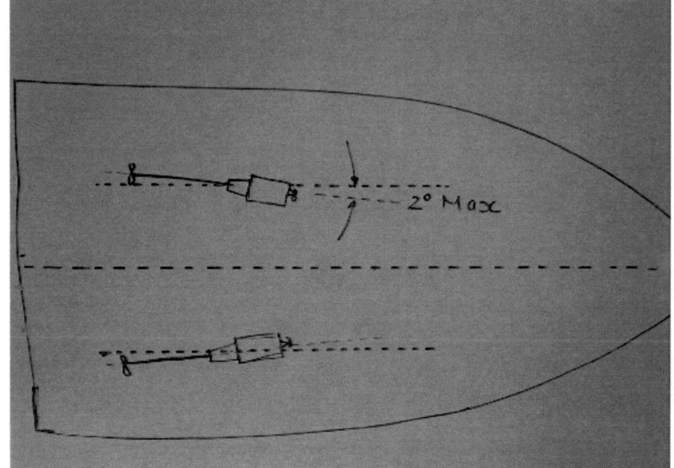

Marine Engine Torque Vs. Marine Engine Horsepower

- Most people make the common mistake of focusing on the marine engine horsepower rather than the marine engine torque.
- There is a common saying that "Horsepower sells a boat however Torque is what actually moves it". This could not be closer to the truth! .

Marine Engine Torque Vs. Marine Engine Horsepower

- One should realize that horsepower is really a measure of the torque over a given period of time .
- This taken into account by the rpm variable in the specification. The following equation may help to shed some light as well.
  - Torque = Hp x 5252 / RPM   (5252 is a constant)

Drive Shaft

In the event your present engine installation transmits significant noise and vibration to the hull, you may wish to consider installing a drive shaft system that incorporates both thrust bearing and CV joints as a part of your re-power project

The Aquadrive system

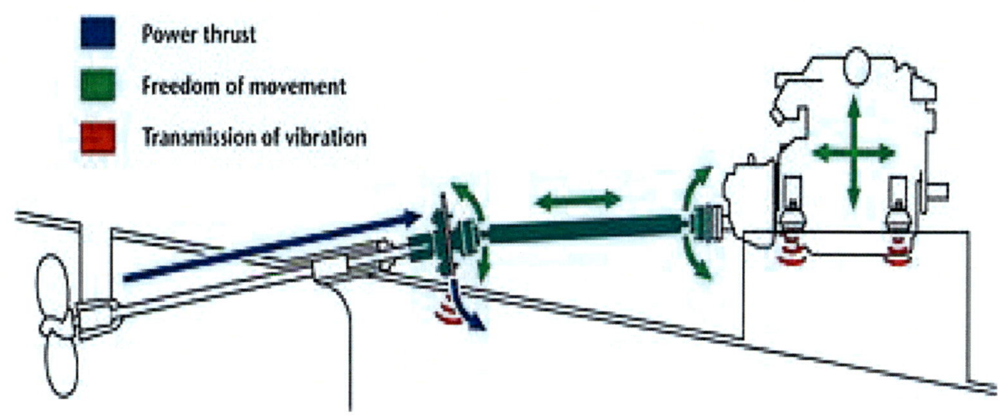

Gear Ratio

Gear ratio is a number, usually expressed as a decimal fraction, representing how many turns of the input shaft cause one revolution of the output shaft

It can be defined as the ratio between numbers of teeth on the meshing gears

In simple gear arrangement, the gear ratio can be simple calculated by looking at the number of teeth on the two gear wheels. It can also be calculated by dividing the tooth count of ring gear to the tooth count of pinion gear, carry out to 2 decimal point

Gear ratio

- **If ratio is too low**:
    - Engine may overwork trying to turn too large a prop at too high a speed
    - At idle, boat will be too fast
    - Engine could overheat from being too heavily loaded
    - Engine may smoke
- **If ratio is too high**:
    - Prop will turn too slowly
    - Boat will not get up to speed
    - Engine will consume too much fuel from over-revving
- **Prop, engine and transmission must be perfectly matched**

Chapter 7
Lower Units & Sterndrives
Service and Maintenance

Out-board Transmissions

- An outboard engine in bolted onto the transom of a vessel.
- They are easier to replaced and can be lifted out of the water when not in use. The older outboards are two stroke, some of the new ones are four stroke engines .

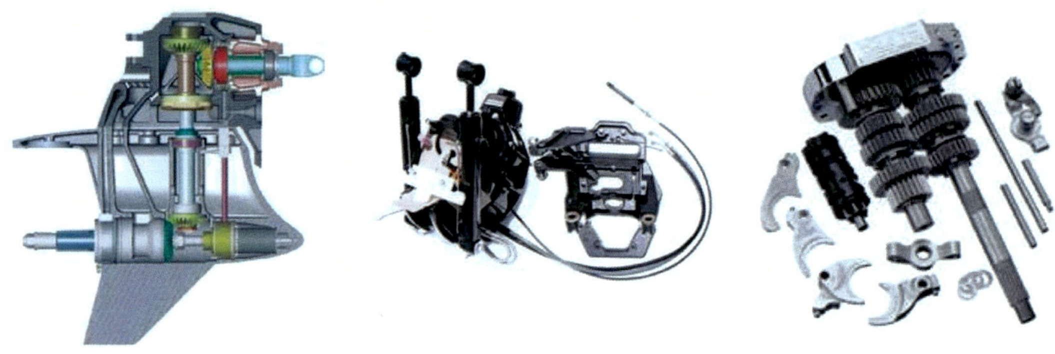

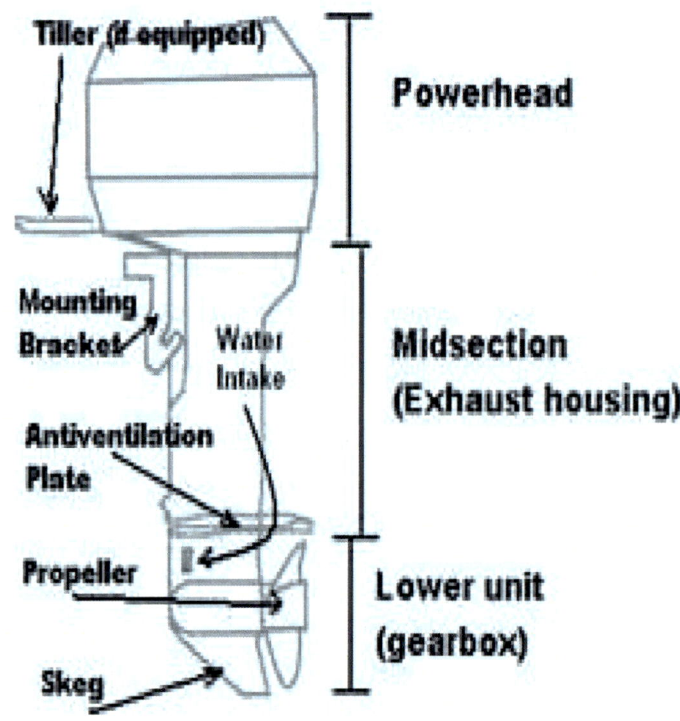

Refrigerant Vs. Salt Water

- Outboard engines are refrigerated by salt water instead of coolant.
- In board engines use raw water strainer between the seacook and the heat exchanger. Outboard engines do not use it

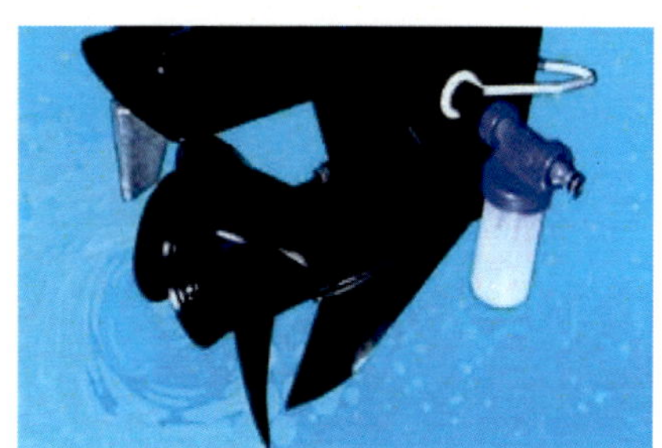

Refrigerant Vs. Salt Water

For this reason the Outboard engines are exposed to blockages in the refrigeration cavities

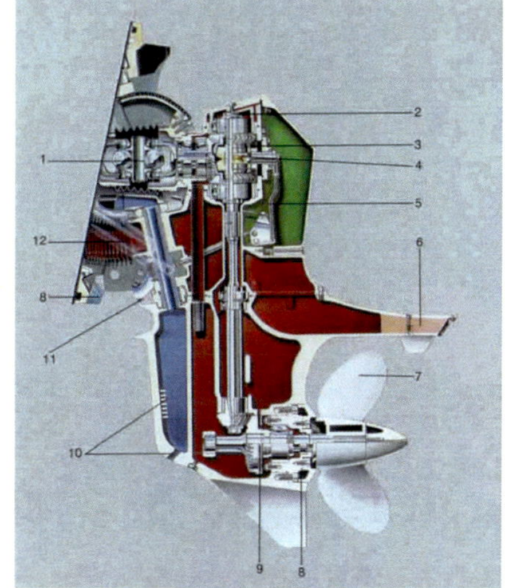

Refrigerant Vs. Salt Water

- Calcium plaques and silica rocks are created into the refrigeration cavities due to the permanent contact between aluminum and salt water

- The heat dissipation is reduced and the temperature block increase quickly

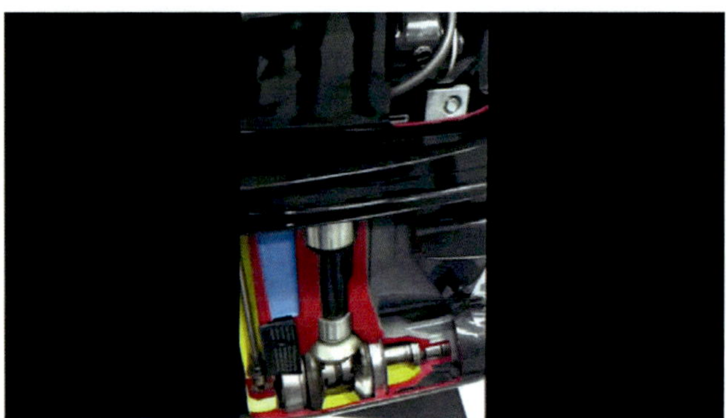

Outboard Cooling System

- The water comes into the outboard's engine through the lower unit, which is below water level as the boat is moving

- There are three main parts to the pump:

- Bottom plate

- Impeller

- Housing

Water Pump

- One rule of thumb is: if the engine overheats and the water pump impeller doesn't look like it just came out of the box, replace it --- then should the problem still persist, go further into the system as needed.

- The driveshaft may have a O-Ring around it on the splines that engage the motor crankshaft.

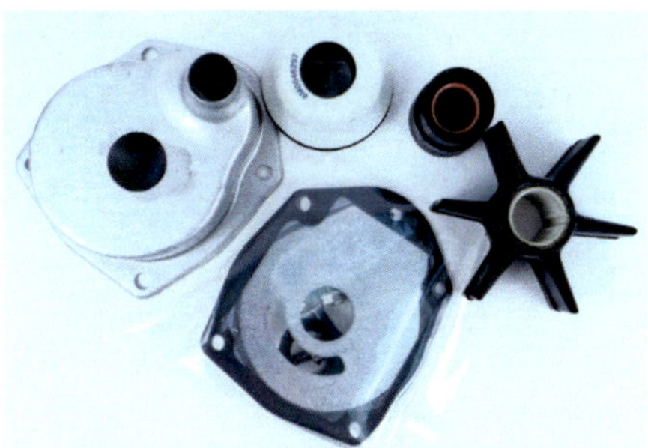

Operation

- If the impeller were exactly in the center of the housing, it would turn with the drive shaft, spinning water within the housing, but it would not pump water

- However, it is designed with an offset impeller

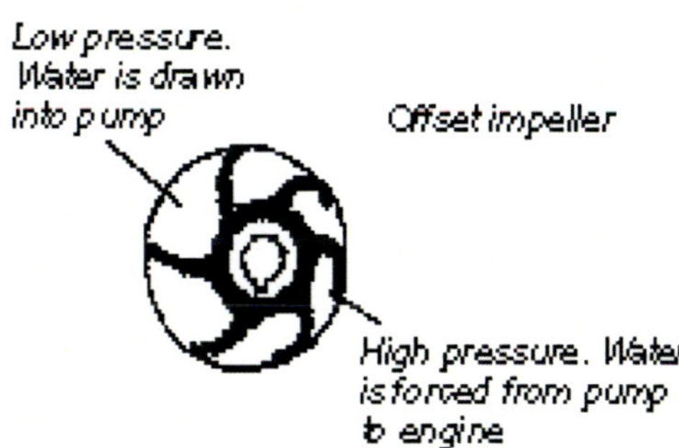

Water Pump

- Things to look for after the water pump is removed are, the impeller can take on a curl rather than have the blades snap straight outwards when removed from the housing.

- if you do not have water coming out of this indicator hole, **& YOU ARE SURE THAT THERE ARE NO BLOCKAGES, your water pump IS NOT WORKING**.

Operation

When the impeller turns, it has a lot of room on one side of the pump, and not much room on the other. As the drive shaft and impeller spin, water enters on the side that has a lot of room, and is pressured out on the side that doesn't have much room.

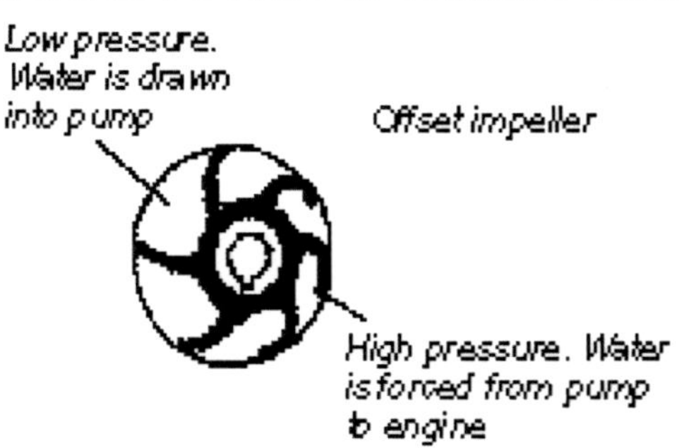

Water Pump

- If the pump housing is the older pot metal as used on the very early motors, replace the whole pump unit with the newer nylon housing kit which has a stainless steel liner that fits inside the housing.

- In most cases you will never need to replace the water pump assembly, just the impeller. Sometimes the rubber impeller becomes un-bonded to the metal hub.

Thermostat

- If a small stone gets jammed in the thermostat, it can be stuck closed which causes the engine to overheat, or it can be stuck open, making warmup very difficult

- If a motor overheats from cooling system failure, quickly remove the spark plugs and pour oil into the cylinder to keep the piston rings from ceasing up in the cylinder

Water Circulation Indicator

- On the side of most motors there is a small hole that emits a stream of water. This is only an indicator that the water pump is working

- During cold weather, the indicator might freeze even though the pump is working well

Raw Water Screen

A screen protects the opening to the water pump on the lower unit. It keeps grass, sticks, and small stones from plugging the cooling system, and destroying the engine

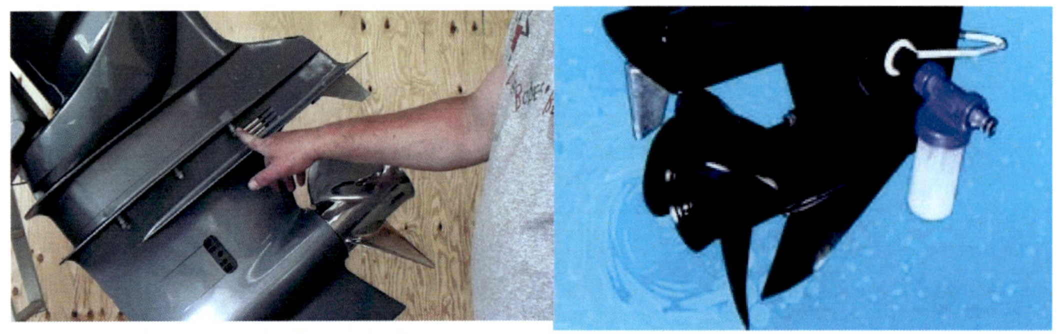

Water Pumps Common Problems

- The most common problems with water pumps are .
  - Worn impellers
  - Worn housings
  - Bottom plates that are worn rough and thin by silt .

- Some people bail water on their engines to cool them until they get home .
- This cools the outer jacket of the engine, but doesn't cool the cylinder walls well at all. The operator might get home, but not without internal damage to the engine

Oil Change

- First, be sure your outboard is in a vertical and upright position.
- Now locate and identify the upper and lower Vent & Drain Screws.

Oil Change

- Remove the BOTTOM drain screw first (not the top).
- If water starts coming out or the oil is really milky in color you have a problem and need to bring your motor to the shop.
- Next, remove the TOP vent screw. This will release the vacuum that exists and allow the oil to drain freely from the lower unit

Oil Change

- The oil should also not appear to be milky. Excessive water in the lower unit could cause it to crack in freezing temperatures.
- To refill the lower unit, insert the bottle/tube into the BOTTOM drain hole.
- Continue adding oil until the lubricant is flowing freely from the top hole and is free from air bubbles.

Oil Change

- Keep the bottle/tube in the bottom hole (or plug the hole with your finger) and re-install the TOP plug. This will create a slight vacuum that will minimize the oil loss while you reinstall the bottom plug.
- QUICKLY remove the bottle (or your finger) from the bottom hole and install the drain plug.

Water & Gear Oil mixed

Take a sample of the oil; send it to the laboratory and analyze the results

Water and Gear Oil mixed

- A gearbox in good repair that's completely full of oil cannot possibly take-on half of it's volume in water.
- If water can get in oil can get out.
- The base oil, synth, mineral or what ever is going to draw down from gravity, and depending on the additives used, they will either protect or leave the surface exposed to rust due to outside moisture accumulated from shut down.

Water and Gear Oil mixed

- A lower unit should be water tight, or fairly close, if it's in good repair.
- water is exposed to metal surface, depending on the type of alloy, some gear oils will not help keep that surface from being damaged due to water intrusion.
- Look in a junk yard and look at some of the pumpkins ring and pinion and see how many have rusted pit marks where water and air had entrained into the gear oil.

Water and Gear Oil mixed

- A lower unit should be water tight, or fairly close, if it's in good repair.

- water is exposed to metal surface, depending on the type of alloy, some gear oils will not help keep that surface from being damaged due to water intrusion.

- Look in a junk yard and look at some of the pumpkins ring and pinion and see how many have rusted pit marks where water and air had entrained into the gear oil.

Gearbox Lubrication Tips

- A bath system is not recommended for either low or high speeds in gearboxes.

- Under low speeds, a low-viscosity lubricant drips off the lower gear before it can be distributed to upper gears. In high speeds, the lubricant can be slung off.

- When selecting a gear lubricant it is important to consider the load,speed, temperature, gear type and finish, and application method.

Gearbox Lubrication Tips

- In general, higher viscosity fluids are needed for higher loads and temperatures, lower speeds, rougher finishes and for worm gears.

- Extreme pressure additives should be used for heavy loads and moderate temperatures, but are not effective with yellow metals such as bronze or brass.

Gearbox Lubrication Tips

- It is generally better to use an oil one grade too high in viscosity than too low in viscosity in gear applications because the more viscous oil will provide more load-carrying ability and maintain a better film strength.

- Refitting gearbox vents with breathers that restrict the ingestion of airborne dirt and debris will help control contamination from entering the unit.

Gearbox Lubrication Tips

- Overfilling a gearbox sump can be just as damaging as under-filling. Overfilling may cause air entrainment and foam, overheated oil and leakage due to overflow.

- Over time, oxidation may occur due to increased temperatures and exposure to air.

Gear Ratio

- Gear ratio is a number, usually expressed as a decimal fraction, representing how many turns of the input shaft cause one revolution of the output shaft.

- It can be defined as the ratio between numbers of teeth on the meshing gears.

- In simple gear arrangement, the gear ratio can be simple calculated by looking at the number of teeth on the two gear wheels. It can also be calculated by dividing the tooth count of ring gear to the tooth count of pinion gear, carry out to 2 decimal point.

Lower unit Gear Ratio

- In a bicycle if the front gear is smaller than the back gear then the cycle is said to have power ratio and the cycle moves easily up the hill.

- if the front gear is larger than the back gear then it has speed ratio and this enables the cycle to go downhill easily.

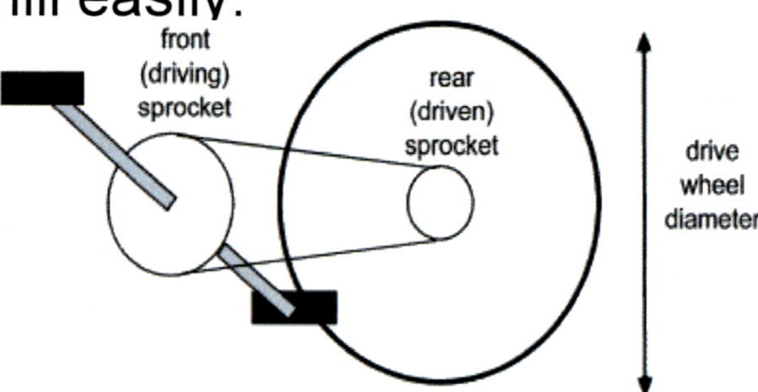

Lower unit Gear Ratio

- On cars with manual transmission more torque application is used, generally a ratio of 3.5:1 is used. Small engine cars and trucks use a final drive ratio of 4.5:1 to provide more torque to enable them to pull or move heavy loads.

- A final drive ratio of 2.8:1 is commonly used, especially with cars with automatic transmission. This means that the drive pinion must rotate 2.8 times to make the ring gear rotate one time

- A drive ratio of 2.1:1 or 2:1 is commonly used in inboard marine transmissions and between 1.4 and 1.95 for outboards

Bevel Gear

- A bevel gear is shaped like a <u>right circular cone</u> with most of its tip cut off .

- Their shaft axes also intersect at this point, forming an arbitrary non-straight angle between the shafts. The angle between the shafts can be anything except zero or 180 degrees.

Bevel Gear

- Bevel gears with equal numbers of teeth and shaft axes at 90 degrees are called *miter gears* .

- The teeth of a bevel gear may be straight-cut as with spur gears, or they may be cut in a variety of other shapes.

Bevel Gear

- *Spiral bevel gears* have teeth that are both curved along their (the tooth's) length; and set at an angle, analogously to the way helical gear teeth are set at an angle compared to spur gear teeth.

- Straight bevel gears are generally used only at speeds below, 1000 r.p.m.

Typical Gear Ratios

- Replacement drives are available in 4 gear ratios: 1.47, 1.62, 1.81, 1.94

194

How Gear Box Works

- The input shaft is driven from the engine or 'power head'. (This input shaft is called the driveshaft)
- The driveshaft has a gear connected to the bottom of it called the Pinion gear.

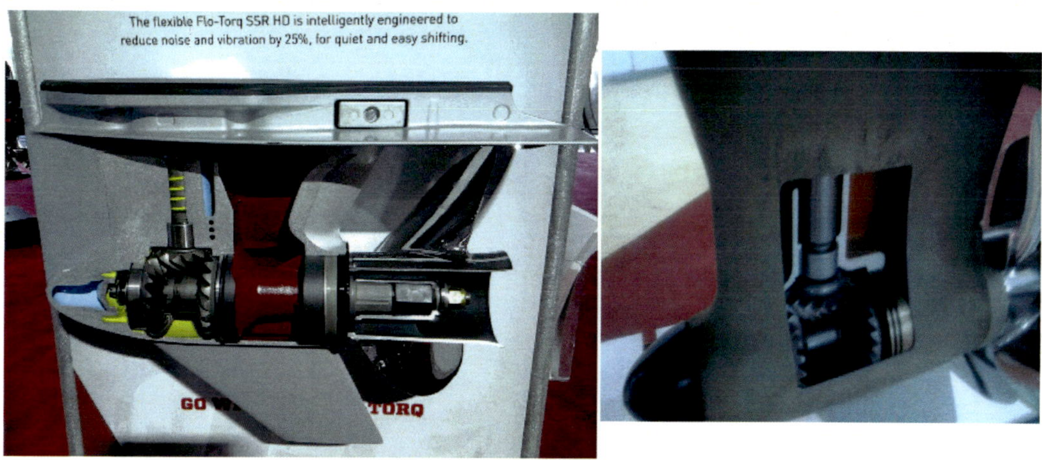

How Gear Box Works

- The pinion gear is always in constant mesh with the forward and reverse gears. When ever the motor is running, the pinion gear is rotating the forward and reverse gears

The Clutch Dog

- Between the forward and reverse gear there is the mechanism that selects gear. This is called the Clutch Dog.

- The Clutch Dog has an internal spline that is in constant mesh with the propeller shaft.

The Clutch Dog

- Picture the Clutch Dog having a front and a back. At each end it has large, square teeth. When you select gear on the motor you move the Clutch Dog forwards and backwards along the propeller shaft

How Gear Box Works

- When the Clutch Dog is moved to select gear it engages with the same sized square teeth on the particular gear that is being engaged.

- Remember that both of the output gears (forward and reverse) are rotating as they are in constant mesh with the pinion gear (input gear).

How to Select Gear

- If you select gear slowly, the Clutch Dog will slowly move into mesh with the gear. This slow movement causes the teeth of the gear to 'skip' over the teeth on the Clutch Dog

- This continuous skipping wears away the nice sharp edges of the teeth and will eventually cause the motor to jump out of gear.

How to Select Gear

- The way to avoid doing damage to your gearbox is to positively select gear. This means to literally bang the engine into gear so you hear one clean engagement of the gearbox, not chatter
- You should also pause in neutral when shifting from forwards to reverse or vice versa. This allows the propeller time to stop spinning

How to select Gear

Do not shift into reverse when you are moving forwards quickly. It creates a huge amount of unnecessary force on the gearbox (and engine)

Slow the boat down of the plane, come back to an idle and pause in neutral to let the boat slow right down. Reverse is purely a reverse, not a break

If you motor is idling higher than around 900rpm in neutral you will be doing damage every time you are selecting gear.

Propeller Diameter and Pitch

A lower blade pitch propeller will push less water in one revolution than one with high pitch, so it is easier for the engine to get it going

Every boat and engine combination has different propeller needs, but in general I have observed that a **higher pitch prop** will give better fuel economy at cruising speeds and a faster top speed, so long as the engine can achieve the rated RPM range for wide open throttle

Pitch Vs RPM

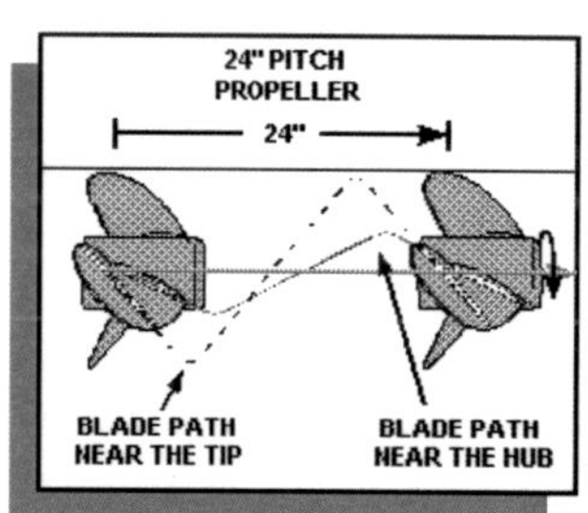

- The pitch is the most important number because different pitches will make the engine R.P.M.'s change at full speed

- The theory is for every one inch in pitch the RPM's will change 150 to 200. The higher the number pitch means lower R.P.M.'s, and the lower the number pitch means higher RPM's

- So the quest is to get the right pitched prop that will get you in your recommended W.O.T. R.P.M. operating range

Diameter and Pitch Recommendation

- A **lower pitch prop** will allow quicker planning, at the cost of some cruise efficiency and top end performance. To find the best prop for your needs, you will need to test a few different ones.

- It is important to conduct a realistic test to properly evaluate potential propellers for your boat. Do not test propellers with the boat empty. Load the boat as you would normally load it when fishing, skiing, or cruising. Don't test it only in flat calm water.

Drive Gear Box

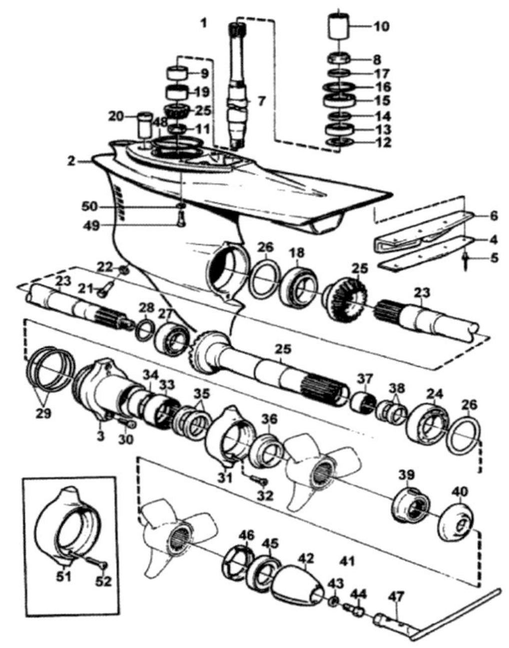

Lower Drive Reparation

- Remove the drain plugs and drain all the oil from the drive.

- Place the gear shifter (remote control) into *forward gear*. The drive must be in forward gear or it will not come off. You will be removing the entire drive

Lower Drive Reparation

Unbolt the hydraulic trim cylinders and let them hang out of the way. Remove the six locknuts and remove the drive from the bell housing

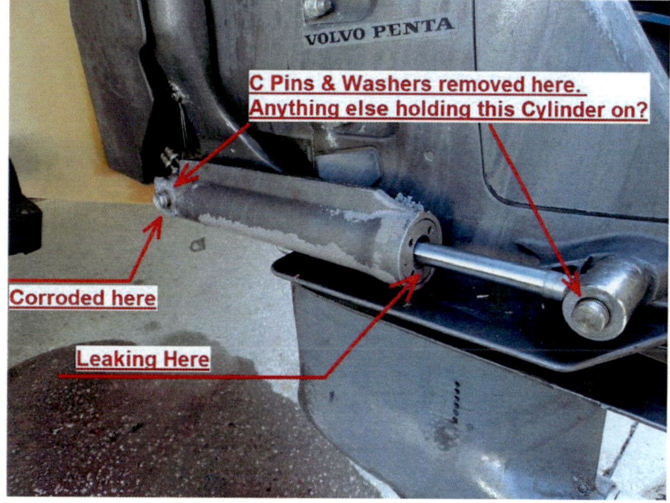

Lower Drive Reparation

- If the drive is stuck, lift it up and let it drop. The "slamming" of the drive dropping will usually break it free.
- Place the entire drive in an adequate drive stand and remove the upper driveshaft housing from the lower gear housing

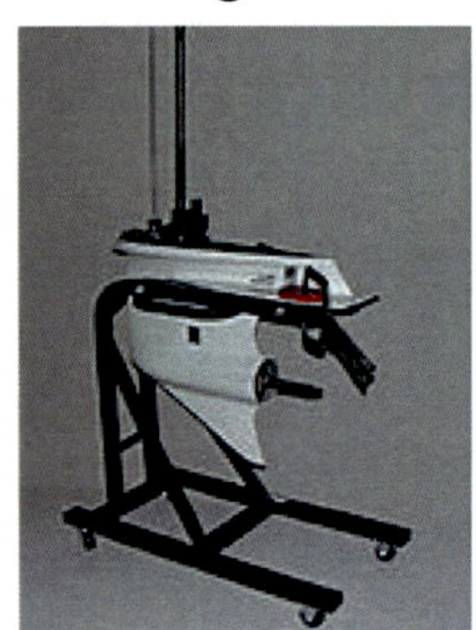

Lower Drive Reparation

- Lift off the upper housing and set it aside.
- Get your "dirty parts tray" nearby as we are going to start setting old dirty parts into it

Lower Drive Reparation

Remove the stainless flat washer that is located on the small splined shift shaft towards the front of the housing. Remove any tubes and slinger seals from on top of the plastic water pump housing.

Remove the Gearcase to Access the Water Pump

If the motor has been used in saltwater to any degree, I will guarantee that at least a few of the bolts will be seized in & will be twisted off

Lower Unit Separation

When you get the lower unit bolts removed, the gear case unit should drop down about 1/2"

Clean Out & Drain is Clear

In the RH photo below, the water pump housing is what you see attached by the 4 bolts. The arrow is pointing to the shifting rod coupler visible on the small shaft

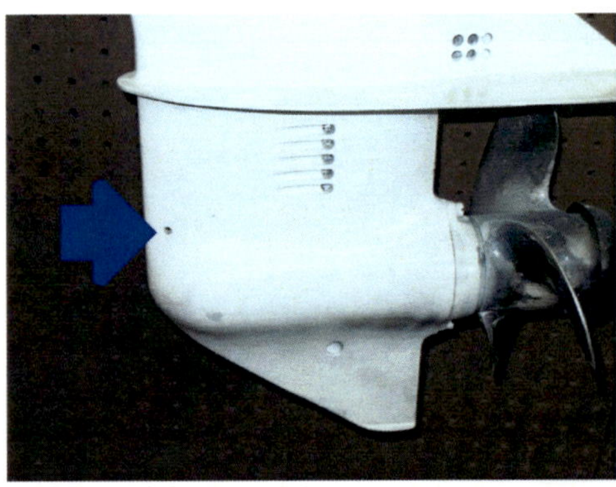

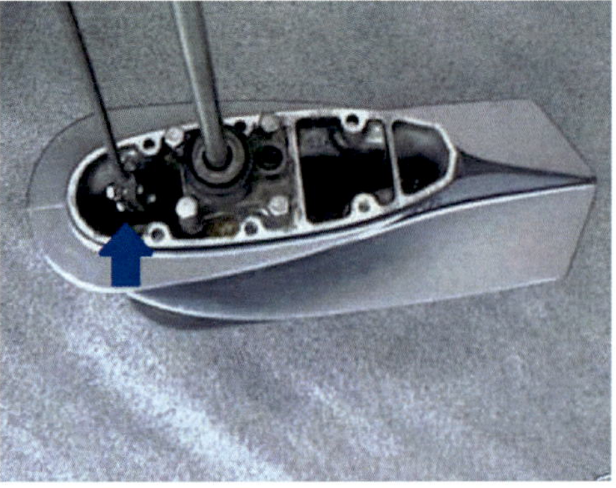

Lower Drive Reparation

Remove the two water pump nuts. Remove the front 1/2" water pump nut. Use a 5/16" socket on a long extension and remove the rear water pump housing screw.

Water Pump Removal

- Lift off the top water pump housing, impeller and key.
- Use two large slotted screw drivers and pry up from underneath the lower water pump base housing and remove it. At this point you should be able to see the upper driveshaft bearing retainer ring. This retainer ring is threaded into the housing but do **not** remove it YET.

Lower Drive Reparation

Use a large slotted screw driver and a hammer and bend straight the locking tabs that lock the large diameter threaded cover nut into the housing

Lower Drive Reparation

- Usually the cover nut is corroded into the threads of the housing. It can be very difficult, if not impossible, to remove.

- The most common (and easiest) method is to use a large impact wrench. Shops usually use a large 3/4 drive impact wrench

Bearing Housing

If the bearing housing is frozen into the gearcase, it will usually be at the rear as the housing slides into the gearcase about 1/2"

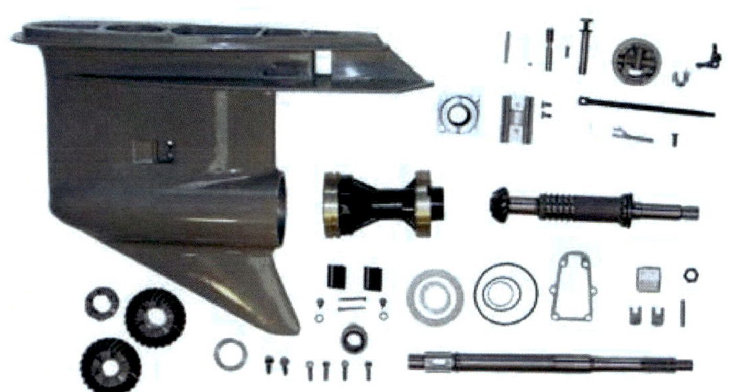

Bearing Housing

If the motor was used in saltwater & not flushed good, there may be salt corrosion between the 2 parts enough to seize them together. If this is the situation you may be able to see where the gearcase has cracked in this area because of the corrosion

Internal corrosion cracks at rear of housing

There is a lot of internal pressure here from the corroding salty aluminum. You can try penetrating oil, tapping the areas with a hammer, even trying some heat from a torch,

but in my experience you are wasting your time

Irreparable transmission

- If you do get it apart, the damage is not really reparable
- If this is the case, it is suggested that you leave it alone, refill the gear oil, replace the drain pug seals & run it as is until you can purchase another better used gearcase assembly, & then try to salvage yours out for spare parts.

Lower Drive Reparation

- An acetylene torch can melt the aluminum housing very easily.

- If, after about 1/2 hour of trying the cover nut just wont come out, give up. It's not worth destroying your impact wrench or retainer wrench. Instead, grad your drill and a sharp 1/4" drill bit and drill 4 holes, as shown.

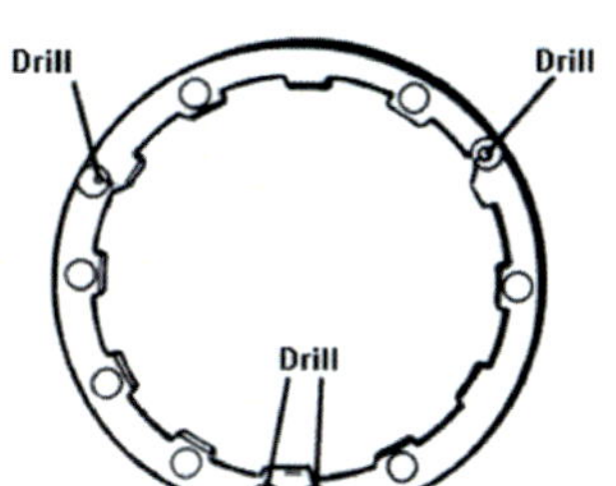

Lower Drive Reparation

Stop and inspect the threads in the housing. This could "make or break" your plan. If the threads are corroded away too badly, the housing will have to be replaced

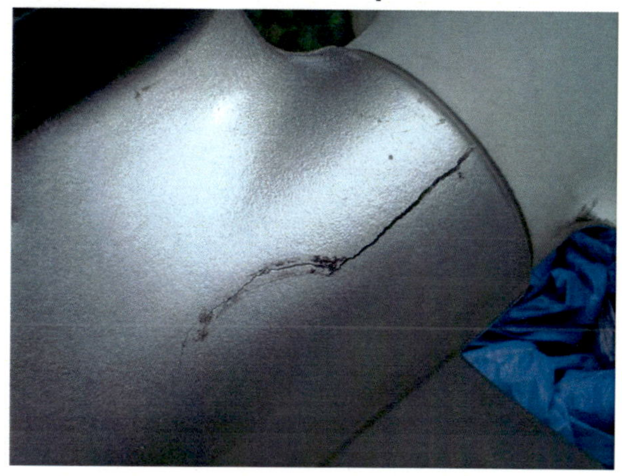

Lower Drive Reparation

The cover nut will strip out and the propshaft bearing carrier will not be held into the case. If the threads look really eaten away, you are having a bad day.

Disassemble or Rebuild the Gearcase

The lower unit, or gearcase needs to be taken off as if you were replacing the water pump impeller. Remove the (2), 1/4" screws that come in from the rear above & below the propshaft exhaust opening

Removal & Replacement of Shaft Seals

To remove the prop shaft seals, you can make a puller by using a 5/16" or 3/8" rod about 15" long

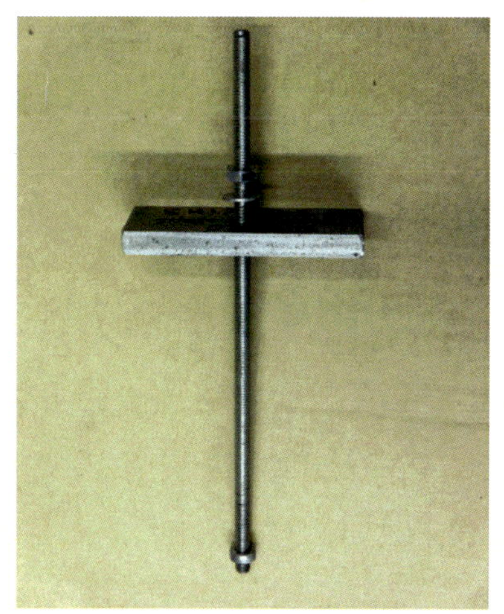

Lower Thrust Mounts Worn

There are 3 squarish rubber mounts, one on each side that are called lateral mounts & a front one that is the actual thrust mount. The side mounts do not normally wear, but the center front one is the one that takes the pressure

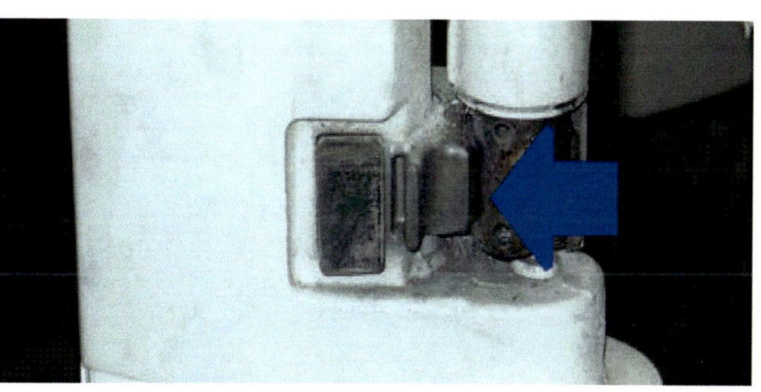

Lower Thrust Mounts Worn

Replacing the front mount is possible without dropping the lower unit ,If you make a tapered wedge out of wood, & drive it between the pivot housing & the midsection

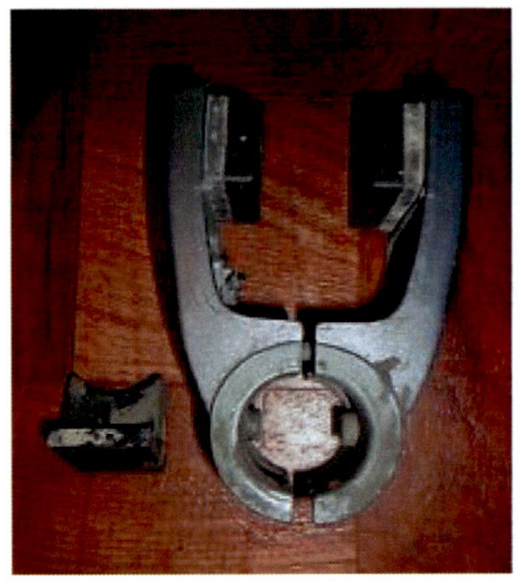 

Hydro-Tail Performance Stabilizer

The Hydro-Tail is excellent for stabilizing boats, reducing cavitations and improving performance for water skiing, fishing or general boating.

- Easy installation on Outboard or I/O Motors
- One piece design
- Great value in hydrofoil stabilizers

For Inboard & Outboard

Transmission Components

- The main components in a boat transmission are the : The bearings, The Gears and The Fluid
- The Bearing functions are
  - Decrease friction, heat and wear
  - Support static weight of shafts and machinery
  - Support radial and thrust loads
  - Allow tighter fitting tolerances

Types of Bearings

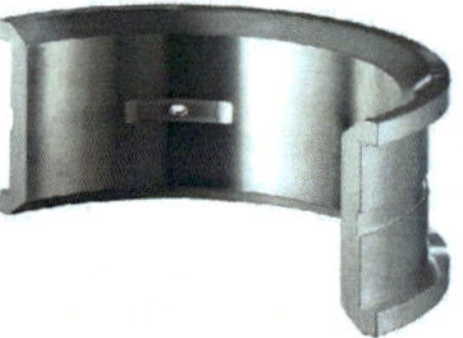

- Two main types of bearings:
  - Plain Bearings
    - Sliding bearings, journal bearings
    - Shafts run directly on the bearing surface
  - Anti-friction
    - Rolling element bearings, rolling contact bearings
    - Balls or rollers as part of the bearing

Load Bearings

- The bearings are also classified according to the load on them

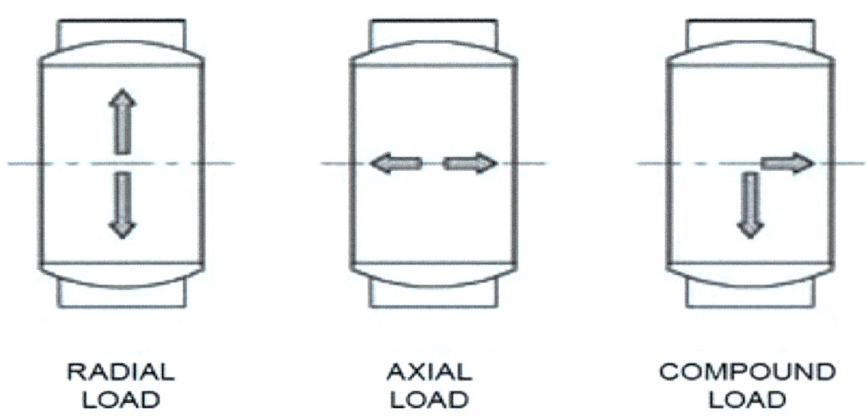

Load Bearings

The gear box in a boat propulsion system works like a fuse in an electric circuit , protecting the main engine of over loads from the propeller and shaft

Bearings and Gears

The life of Bearings and gears depends of :
- The lubricant quality
- The engine alignment
- The propeller selection
- The operating temperature of the oil
- The proper driving of the boat
- The proper driving of the boat
- The Bonding System

Radial ball Bearings

Radial ball bearings tolerate relatively high-speed operation under a range of load conditions. Bearings consist of an inner and outer ring with a cage containing a complement of precision balls

Radial Ball Bearings

The input and output shaft works over some ball and compound bearings that support heavy loads

100 % Radial Bearing

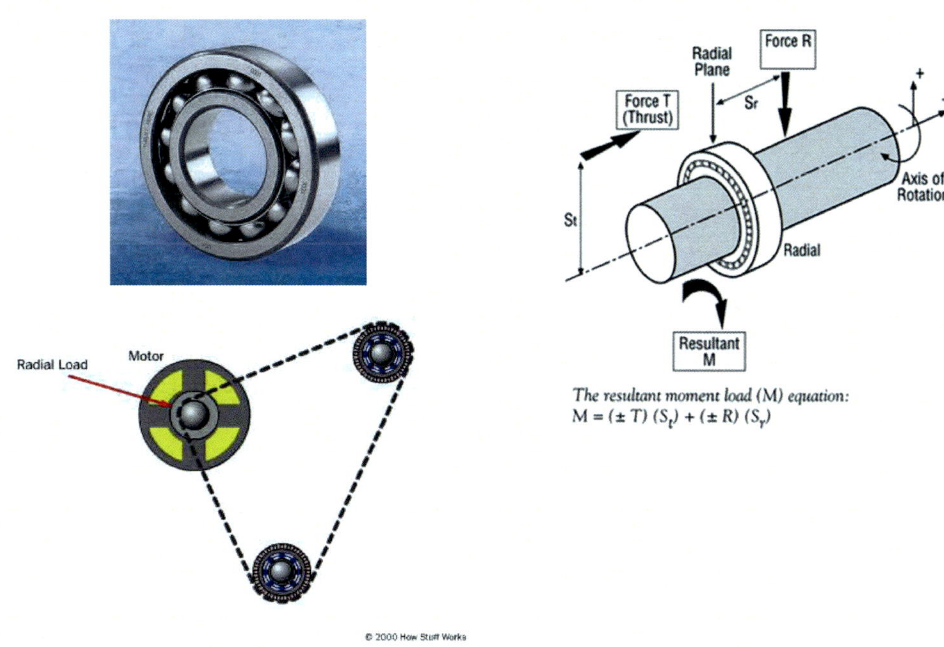

The resultant moment load (M) equation:

$$M = (\pm T)(S_t) + (\pm R)(S_r)$$

100 % Axial Load

Compound Load

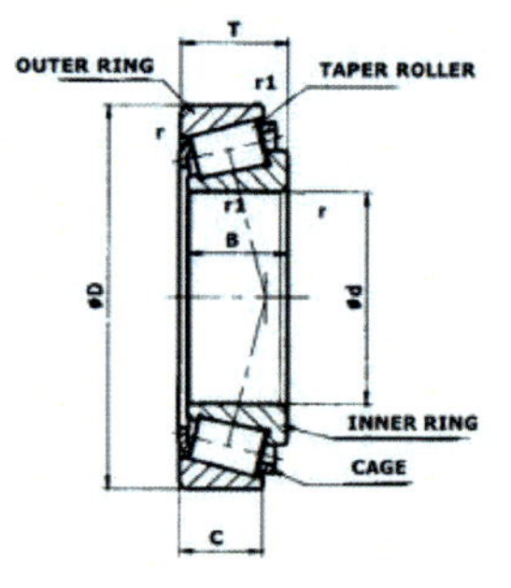

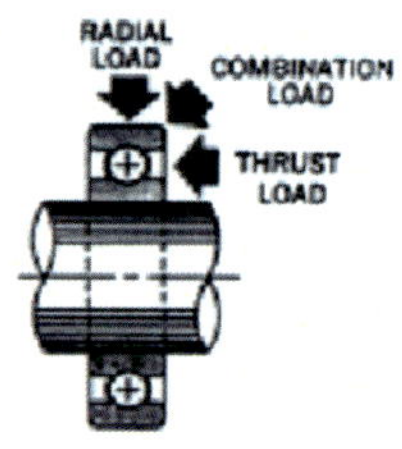

Angular ball bearing

Angular contact ball bearings are designed for combination radial and axial loading and are used in stern-drive shafts and V-drive transmissions.

Angular ball bearing

- Angular contact bearings are designed such that a contact angle between the races and the balls is formed when the bearing is in use.
- The major design characteristic of this type of bearing is that one or both or the ring races have one shoulder higher than the other

Roller bearings

- Three basic types:
  - Straight roller bearings
  - Needle roller bearings
  - Tapered roller bearings

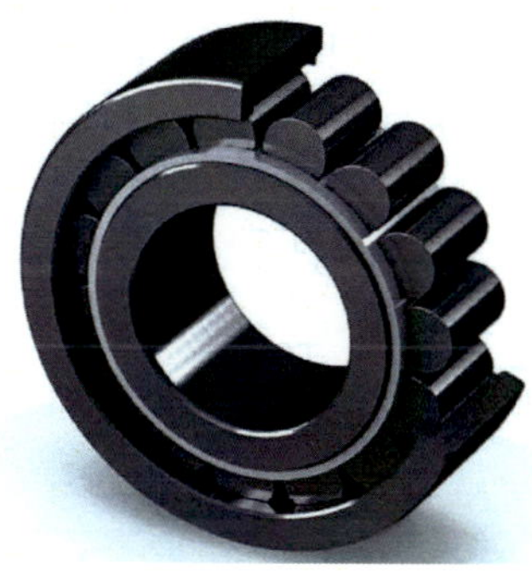

Roller Bearings

Common roller bearings use cylinders of slightly greater length than diameter. Roller bearings typically have higher load capacity than ball bearings, but a lower capacity and higher friction under loads perpendicular to the primary supported direction

Needle Roller Bearings

Needle roller bearings use very long and thin cylinders. Often the ends of the rollers taper to points, and these are used to keep the rollers captive, or they may be hemispherical and not captive but held by the shaft itself or a similar arrangement

Needle Roller Bearing uses

Since the rollers are thin, the outside diameter of the bearing is only slightly larger than the hole in the middle. However, the small-diameter rollers must bend sharply where they contact the races, and thus the bearing [fatigues](#) relatively quickly

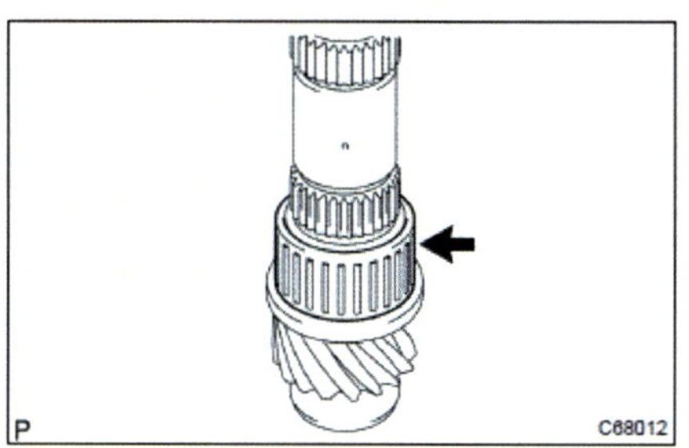

Spherical Roller

Spherical roller bearings use rollers that are thicker in the middle and thinner at the ends; the race is shaped to match. Spherical roller bearings can thus adjust to support misaligned loads

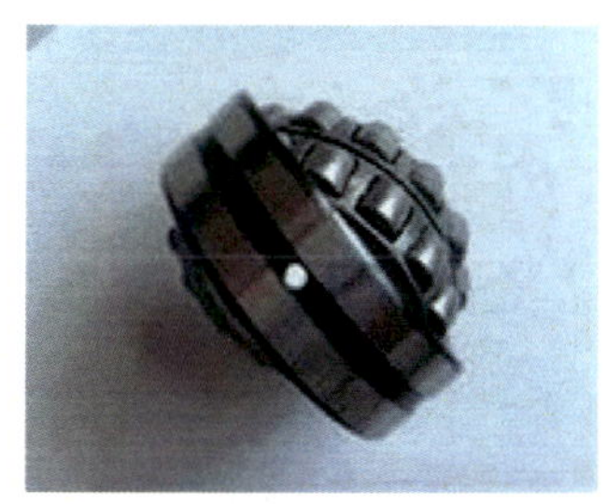

Thrust Bearings

Thrust bearings are used to support axial loads, such as vertical shafts. Commonly [spherical](), conical or cylindrical rollers are used; but non-rolling element bearings such as hydrostatic or magnetic bearings see some use where particularly heavy loads or low friction is needed

Thrust Bearings

Thrust bearings are used to support axial loads, such as vertical shafts. Commonly [spherical](), conical or cylindrical rollers are used; but non-rolling element bearings such as hydrostatic or magnetic bearings see some use where particularly heavy loads or low friction is needed

Plain Bearings

- By far, the most common bearing is the <u>plain bearing</u>, a bearing which uses surfaces in rubbing contact, often with a <u>lubricant</u> such as oil or graphite.

- With suitable lubrication, plain bearings often give entirely acceptable accuracy, life, and friction at minimal cost. Therefore, they are very widely used.

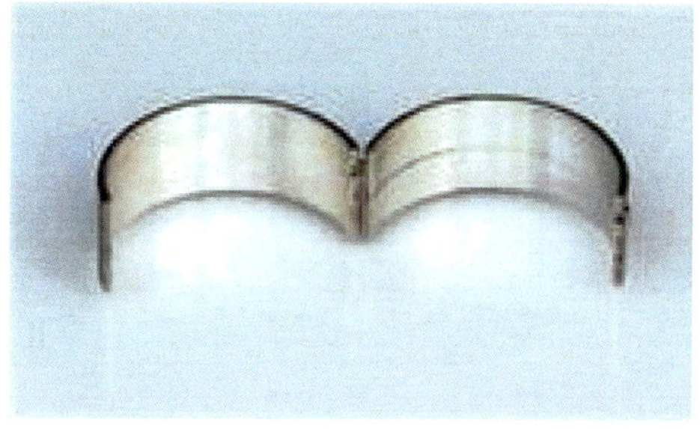

Plain Bearings

- In a plain bearing operating under hydrodynamic or full-film lubrication, a film of lubricant completely separates the shaft and bearing .

- In many situations, the bearing itself contains or acts as the lubricant (Such pre-lubricated or self-lubricating bearings) .

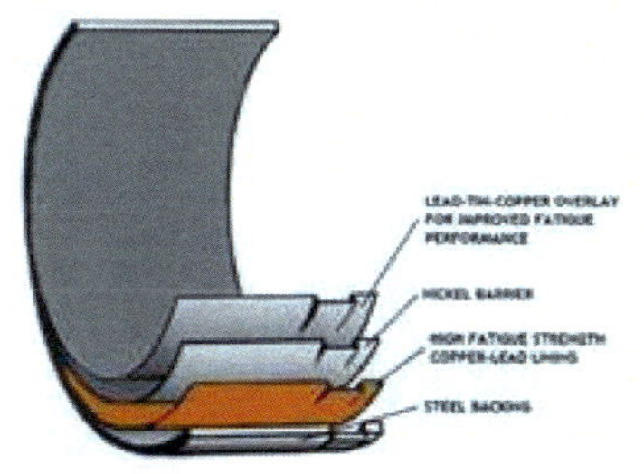

Bearing Material

- Bronze has probably been the most familiar plain bearing material because a variety of characteristics can be imparted to it by adding other metals

- In general, softer materials are designed for lighter loads and higher speeds; harder materials for higher loads and lower speeds

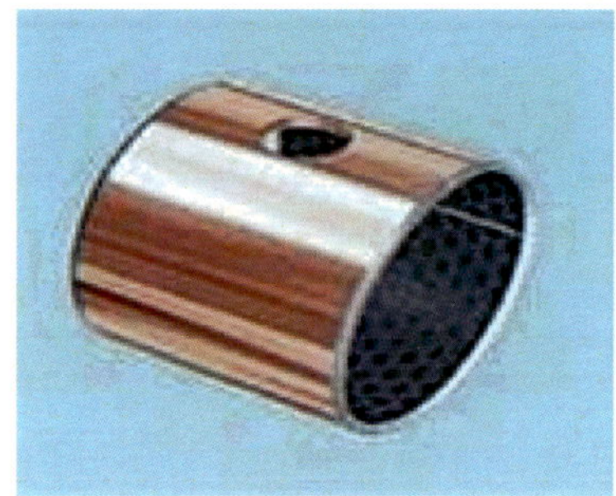

Babbitt Bearings

- The softest metallic bearing materials are babbitts. Both **tin** and **lead**-based babbitts have been widely used as bearing materials for years.

- They are much softer than bronze and are able to embed foreign particles, which helps prevent shaft scoring or wearing

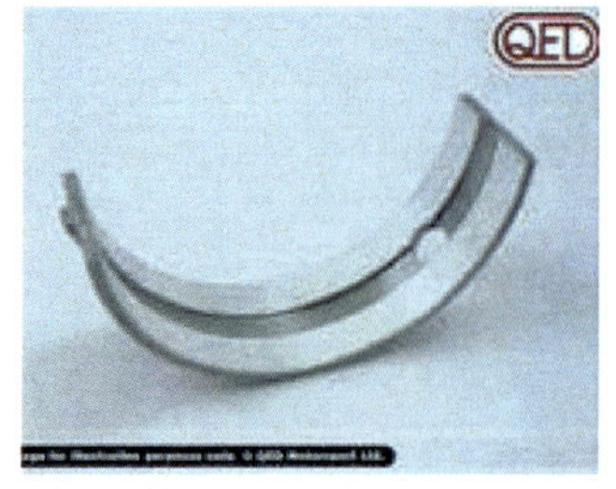

Babbitt Bearings

- Babbitt bearings offer excellent resistance to shaft scoring and seizing in boundary lubrication conditions .

- Copper-lead is also soft, though it approaches some of the softer bronzes in hardness .

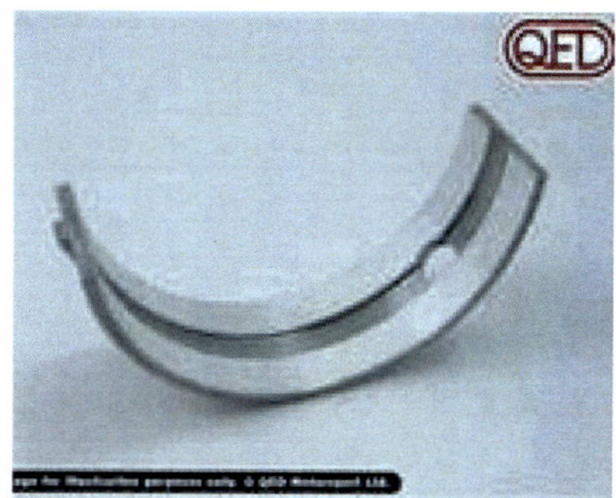

Bronze- Lead Alloys

- High lead content helps these bronzes resist seizing or scoring of the shaft. Maximum operating temperature of leaded bronzes runs typically from 400 to 450°F

- Decreasing the lead concentration increases strength and hardness of the material, but decreases its conformability, scoring resistance, and ability to embed foreign particles

Seals & Gaskets

Gaskets

- A **gasket** is a mechanical seal which fills the space between two or more mating surfaces, generally to prevent leakage from or into the joined objects while under compression

- Gaskets are commonly produced by cutting from sheet materials, such as gasket paper , rubber, silicone, metal, cork, felt, neoprene, nitrile and rubber

Gasket Material

- The required compression for your gasket material will depend many factors including
    - 1)Surface area
    - 2)Pressure being sealed
    - 3)Size of bolts (assuming bolts are being used)
    - 4)Number of bolts
    - 5)Condition of the bolts
    - 6)Lubrication on the bolts

Gasket Material

- Gaskets are commonly produced by cutting from sheet gasket materials, such as:
- Gasket Paper, Non-asbestos, Rubber, EPDM, Nitrile, Buna, Neoprene, Flexible Graphite, Grafoil, Aflas, Kalrez, Viton, Silicone, Metal, Mica, Felt or a plastic polymer such as PTFE, Peek, Urethane, or Ethylene Propylene (EP)

Gaskets & Sealant

- It is usually desirable that the gasket be made from a material that is to some degree yielding such that it is able to deform and tightly fills the space it is designed for, including any slight irregularities .

- A few gaskets require an application of sealant directly to the gasket surface to function properly

Metal Gaskets

Some (piping) gaskets and Exhaust Manifold gaskets are made entirely of metal and rely on a seating surface to accomplish the seal; the metal's own spring characteristics are utilized (up to but not passing σ_y, the material's yield strength).

Hydraulic Seals

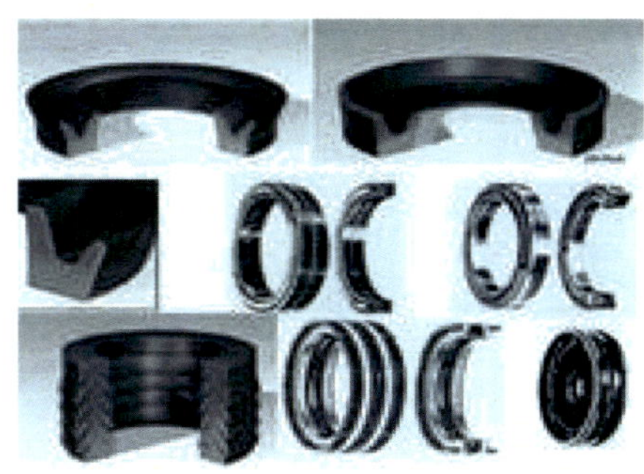

Uses, Applications and Installation Tips

Hydraulic Seals

Hydraulic seals are devices that are designed to keep fluids from escaping from engines, transmissions, cylinders or pumps, and to prevent foreign contaminants from entering them

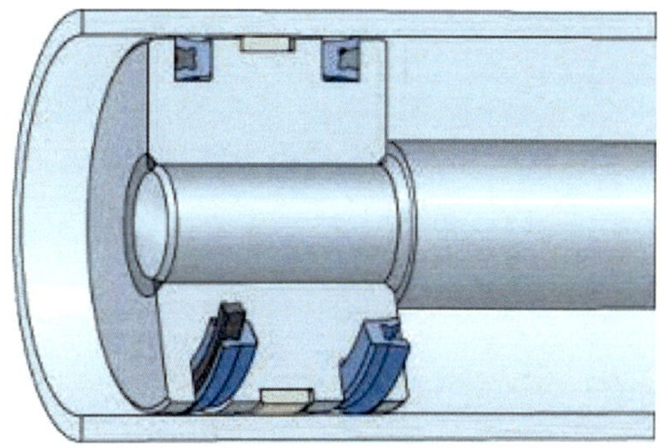

Hydraulic Seals

- A piece of material or a method that prevents or decreases the flow of fluid or air between two surfaces
- The sealed surface may be stationary or have movement between them

Standard Loaded Lip Seal

Loaded Lip Seals are very versatile and can be used as rod seals or piston seals. They are considered to be multi-purpose hydraulic seals and are designed for significantly improved performance

Rotary Shaft Seal

Hydraulic seals include shaft outer diameter **(OD)** or seal inner diameter **(ID),** housing bore diameter or seal outer diameter, axial cross section or thickness **(t),** and radial cross section. Common features for hydraulic seals include spring loaded, integral wiper, and split seal

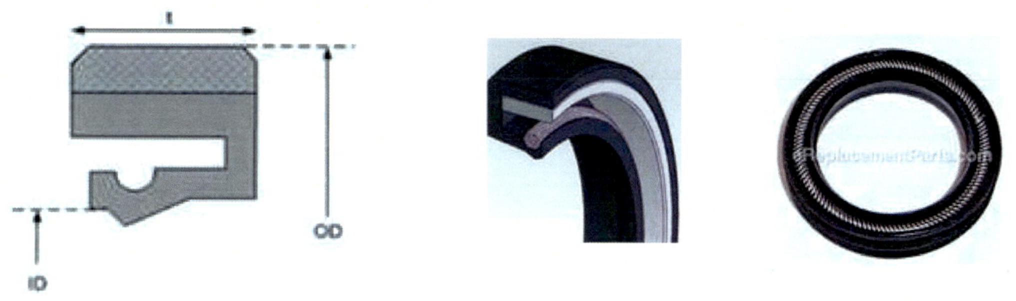

Duties of a Hydraulic Seal

- Prevent lubrication leaks
- Keep out dirt and foreign bodies
- Keep different fluids apart
- Remain flexible
- Seal rough surfaces
- Wear faster than more expensive parts

Types of Hydraulic Seals

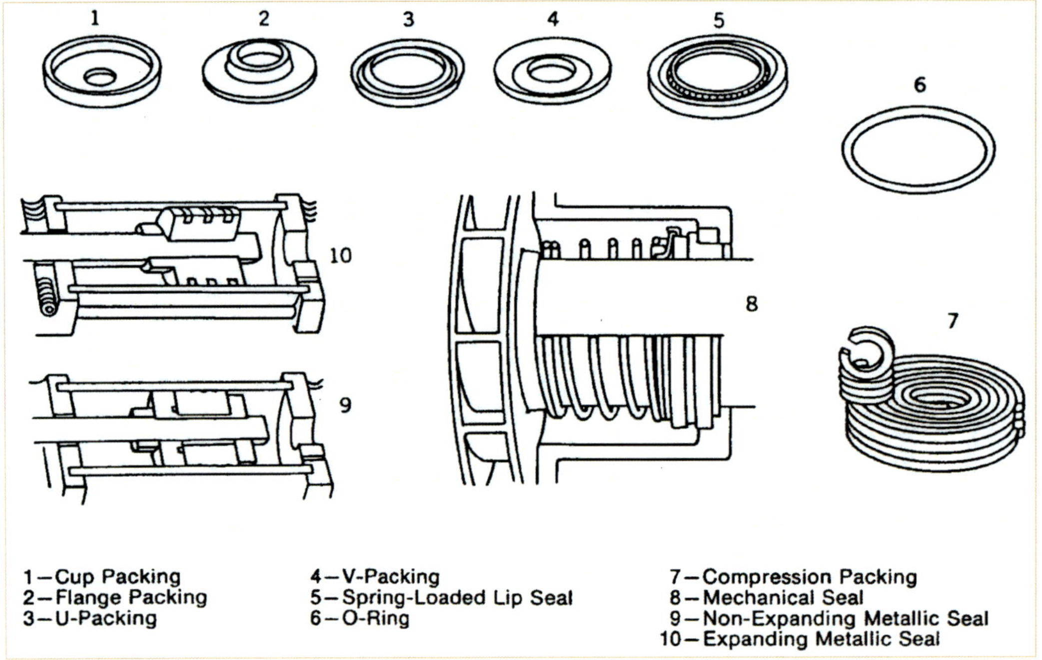

Replacing Oil Seals

- If a seal is removed for whatever reason – it should be replaced
- When installing a seal:
  - Ensure it is not damaged or distorted
  - Install the correct way
  - Lip-type seals must have the lips pointing in the right direction

Checking Seals for Leakage

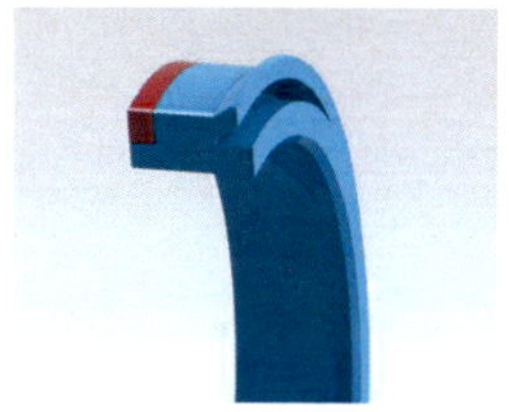

Hydraulic Oil Seal Installation

- Protecting seals while installing
- Lubricating seals before installing
- Checking sealing surfaces
- Checking boots
- Hydraulic brake seals

Installing New Seals

- Install Genuine seals
- Use Proper fluids
- Seals and Fluids clean and free of dirt
- Clean shaft or bore
- Lubricate the seal
- Metal cased seal – lubricate outside diameter

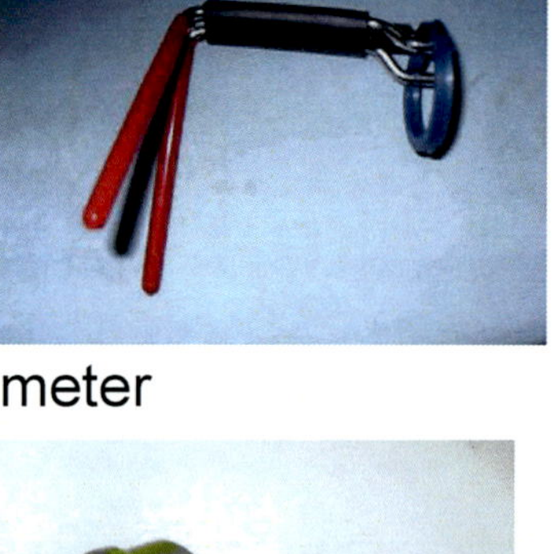

Installing New Seals

- Use factory recommended tools
- No undue force on packings
- Shim stock
- Driven in evenly
- Check unit operation before start up
- Dirt and grit

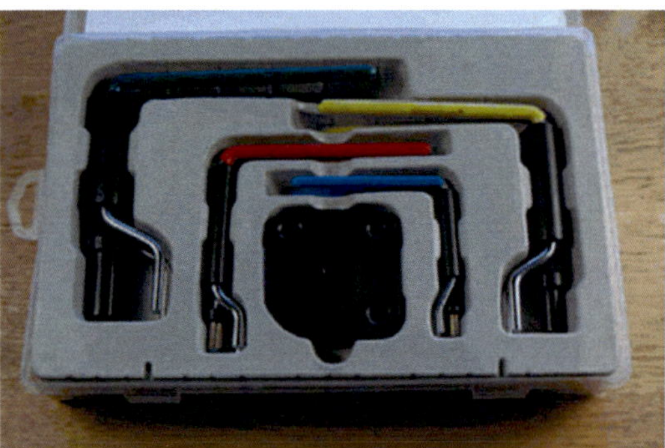

O-Ring Seals

O-Rings or Rotary Seals

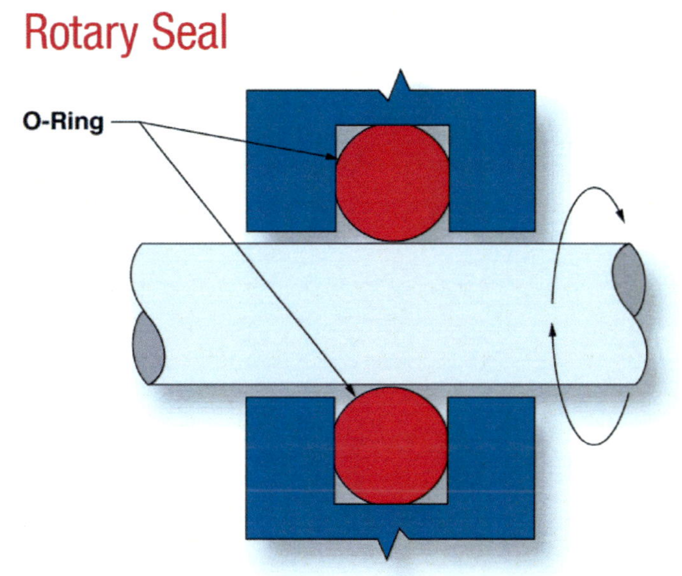

O-Rings Materials

- **Propylene**. It has fair resistance to brake fluids and phosphate esters while exhibiting good resistance to petroleum oils (Diesel Fuel)

- **Ethylene Propylene** . This material was introduced in 1961 and found broad acceptance in application requiring excellent resistance to Skydrol (ATF) and other phosphate ester fluids at higher temperatures.

O-Rings Materials

- **Fluorocarbon**. They are especially good for hard vacuum service (Surgery equipment) and low gas permeability (Fuel Pumps).

- **Fluorosilicone.** They are used in aerospace applications for fuel systems (Fuel Injectors and Fuel Injection Pumps for Diesel Engines).

- **Neoprene**. Due to the excellent resistance to refrigerants such as Freon, neoprene compounds are used in refrigeration systems

How to Specify O-Rings

When you order an O-ring, the manufacturer needs to know the inside diameter (I.D.), the cross sectional diameter (W), and the compound (elastomer formula) from which it is to be made

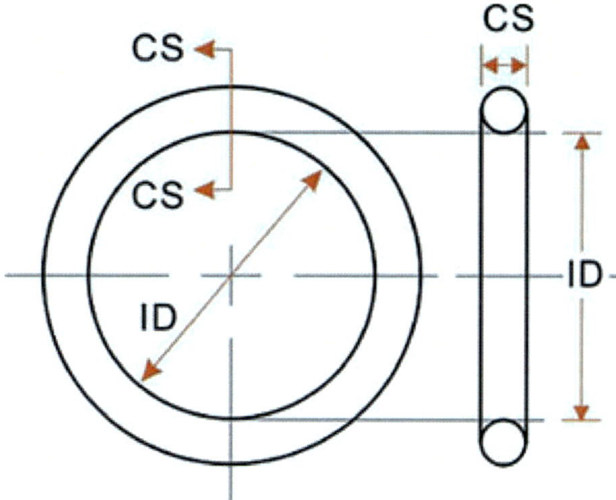

How to Specify O-Rings

Selection of the appropriate O-ring is based on such factors as chemical compatibility, the temperature range of the given application, sealing pressures, lubrication needs and durability

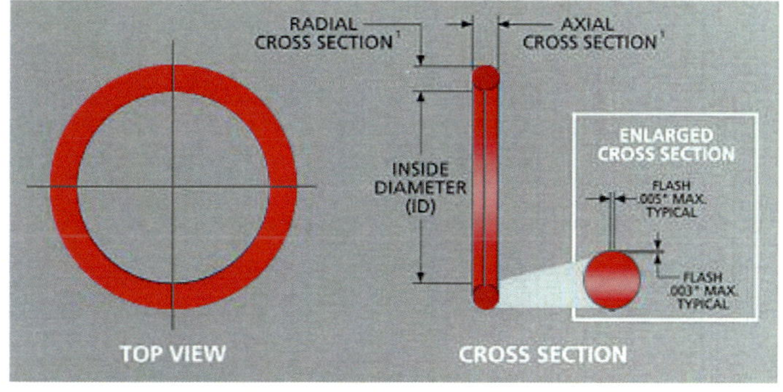

O-Ring Color Codes

Color	Swatch	O-Ring	Color	Swatch	O-Ring
Brown (469 C)	PMS 469		Orange (Orange 21 C)	Orange 021	
Dark Purple (260 C)	PMS 260		Red (Red 032 C)	Red 032	
Grey (422 C)	PMS 422		Tan (465 C)	PMS 465	
Light Blue (Process Blue C)	Process Blue		Teal (322 C)	PMS 322	
Light Purple (676 C)	PMS 676		Yellow (Yellow C)	Yellow	